Interactive Television

McGraw-Hill Visual Technology Series Titles

BUEHRENS • *DataCAD: An Illustrated Tutorial*

HELLER, HELLER • *Multimedia Business Presentations: Customized Applications*

KIWIC, THOMPSON • *Videoconferencing: Design and Implementation*

MACHOVER • *CAD/CAM Handbook*

REYNOLDS, IWINSKI • *Multimedia Training: Developing Technology-Based Systems*

JERN, EARNSHAW, BROWN, VINCE • *Scientific Visualization Graphics*

Interactive Television

A Comprehensive Guide for Multimedia Technologists

Winston William Hodge

McGraw-Hill, Inc.

New York San Francisco Washington, D.C. Auckland Bogotá
Caracas Lisbon London Madrid Mexico City Milan
Montreal New Delhi San Juan Singapore
Sydney Tokyo Toronto

Library of Congress Cataloging-in-Publication Data

Hodge, Winston William.
Interactive television : a comprehensive guide for multimedia technologists / by Winston William Hodge.
p. cm.
Includes index.
ISBN 0-07-029151-9
1. Interactive video.
TK6687.H63 1994
384.55—dc20 94-37454
CIP

1 2 3 4 5 6 7 8 9 0 DOH/DOH 9 9 8 7 6 5 4

ISBN 0-07-029151-9

The sponsoring editor for this book was Roland S. Phelps, the supervising editor was Robert Ostrander, the manuscript editor was Aaron Bittner, and the director of production was Katherine G. Brown. This book was set in ITC Century Light in Blue Ridge Summit, Pa.

Printed and bound by R. R. Donnelley & Sons Company.

To Sue, Karen, and Nancy with all my love.

Contents

Acknowledgments

Interactive television and advanced multimedia could be considered to have been born of early space flight and airplane simulators in the early 1960s. Computer-generated video, created by large mainframe computers, simulated environments for astronauts and pilots. Television projection systems were used as visual display devices and control panels were interfaced to these computers to permit the astronaut or pilot to interact with his simulated environment. Some simulators provided other environmental effects (such as G-Load simulation, pitch, and yaw) by physically controlling the participant's environment. Different flight simulators provided different levels of interactivity. This was also, in effect, early virtual reality.

It is unlikely that one would want an interactive television system that would provide the same degree of reality as provided by flight simulators, in that it would not be necessary to rotate or tilt the viewer's living room or even his chair; but it is likely that interactivity to permit the dynamic selection of data, movies, or specific kinds of news, interaction with TV talk shows or other programs, and the capability to provide near-real-time responses would be desirable.

The systems planned for tomorrow will be controlled regionally by a central facility (head end) and will provide movies or video on demand (VOD), time-shifted TV, interactive news, tele-education, home shopping, video games, and virtual reality as well as conventional television and telephone service. The service providers of these systems will be the cable television (CATV) providers, TELCO, or combination companies.

These systems will permit the user to sit in front of his or her TV, pick up a single-button remote control, point it at the screen to a menu that appears, navigate through myriad programming possibilities, select a movie, order food to be delivered to his or her home for consumption during the selected program, and interact with the program's producers with regard to his or her likes and dislikes. The possibilities are endless.

This book will discuss: the converging of TV, multimedia, and virtual reality; the technology behind video on demand, interactive TV and advanced multimedia; the architectural concepts, approaches, standards, and requirements; differences between true video on demand and near video on demand; communications requirements and storage systems requirements; system control, human interfaces, costs, and the future.

I would appreciate written reader feedback so that subsequent editions of the book can reflect the needs of my readers more precisely. Comments can be directed to me on CompuServe at 74220.2126@compuserve.com (Winston Hodge) or via fax at 714-692-5462.

As the principal author of this book, co-author of some chapters, and sole author of others, I would like to thank all the people who have made this work possible. Because of the timeliness of this subject matter, some of these chapters have been published in various professional journals and presented at professional conferences such as SMPTE and NCTA. Feedback from these presentations has been documented in this book. The CATV, TELCO, computer and communications industries have contributed with plans, drawings, system descriptions, and the like. But mostly, I would like to thank my cocontributors and contributors for their time and dedication to this project. I am truly indebted to them for their contributions.

To Lawrence Taylor, my partner and codeveloper of certain video server and TV set top box technology, contributor of the preface, chapters 3, 5, 6, and 11, my heartfelt thanks and appreciation.

To Stu Mabon, my chapter 3 and *SMPTE Journal* (September 1993) article coauthor, TVOD video server and friendly disk collaborator, deliberator, and friend, who provided valuable support throughout this effort, with hopes we may do more together in the future, thank you.

To Robert Block, president of International Communications Technology Corporation, the founder of Select TV, a brilliant futurist, an individual who permitted me to assist him with his over-the-air broadcast on pay-per-view technology, and coauthor of a paper we jointly delivered to the Society of Motion Picture & Television Engineers (which was only slightly modified to become chapter 13, "The Future"), thank you very much.

To Jack Powers, my CATV and TELCO engineering expert and friend, who helped me write chapters 2 and 14 and provided a good ear and valuable feedback throughout, thanks!

To Chuck Milligan, Director of Engineering at Storage Technology Corporation, who coauthored chapter 7 (which compares true video on demand with near video on demand), thank you for your help, especially in the area of mathematical modeling.

To Bud Junker, my telephone systems engineering expert, who researched demographic and other numerical data to validate concepts and helped reduce it for presentation, who proofed this entire manuscript and found errors others missed, thank you for adding to the credibility of this book.

To Don Dulchinos, Director of Research at Cable Labs, thanks for your help with the valuable information you contributed to this book's preface and chapter 4. These valid and significant CATV statistics demonstrated the reasons why this topic should be covered.

To Roger Pence, Director of Engineering at the NCTA Science and Technology Department, who provided me input from his published writings relevant to chapter 10 (ATM), thank you for your contribution.

To Jim Chiddix, Senior Vice President of Engineering and Technology at Time-Warner, my profound thanks for your advice and copies of your numerous published papers and relevant drawings and charts that found their way into various portions of my book, and for helping me obtain current but preliminary MPEG 2 specifications.

To Mike Demuro, president of Intercor and publisher of the weekly "Interactive TV Report," thanks for the business justification and statistical material you provided, which made its way into this book in various ways, generally as part of my chapter openings.

To David Hench, Hodge Computer Research's (HCR) Managing Director of Imaging Transforms, thank you for assistance in proofing chapter 5 relevant to the various image compression technologies.

To Gordon Smith, Vice President of Technology at HCR, patent holder of 7 image compression patents, digital imaging inventor, contributor to chapter 5, and technical proofreader of the entire book, thanks for your considerable time and effort.

To Aaron Bittner, who put a lot of effort into this book's final edit, thank you for your careful attention.

Much of this undertaking was a partnership dedicated to bringing together under one cover the various ideas that, when combined, would provide one integrated volume on the subject of interactive TV and advanced multimedia. This book is intended to be the single-volume reference manual on this topic, the guide for the ITV/multimedia practitioner and/or enthusiast.

This book was created using distributed computing technology and communications facilities. It was initially prepared on a Macintosh computer, simultaneously on a Local Area Network and a Wide Area Network connected to other MACs and PCs. Word processing was done mostly on Microsoft Word (MACs and PCs), but partially using WordPerfect because this was my partner's preferred word processing software.

Communications between various geographic locations was facilitated using Apple Remote Access and Microphone II Pro. Sometimes the manuscript was sent via CompuServe or the Internet when the recipient was not immediately available. Sometimes the document was faxed, and then (using OCR software) converted back to machine-readable and editable form. As can be seen, we used the tools of our trade to create this document, because most of it was created via telecomputing. A list of software used in the preparation of this book is as follows:

- Microsoft Word (word processing)
- Microsoft Excel (spreadsheet)
- Cricket Graph (graphing)
- Microphone II Pro (modem control)
- Xerox Acutext (OCR)
- FAX STF (fax software)
- Apple Remote Access Server (remote networking)
- Adobe Photoshop (pictorial conversions and editing)

- MacDraw (drawing program)
- XyWrite III (word processing)
- QuarkXpress (desktop publishing)
- Novell Netware 3.12 (local area networking)
- Word for Word (format translation)
- Graphics Workshop (bitmap editing)

and others . . .

And all the above software ran under Macintosh Operating System 7.1, MS-DOS 5.0, MS-DOS 6.2, Novell Netware 2.0, and Windows 3.1.

Needless to say, this was a fun and inspiration-filled project, and I hope the reader attacks this book with the same enthusiasm that I have.

Preface

Highly evolved telecommunications networks, interconnected and delivering telecommunications and data services around the country and the world, will enable the delivery of virtually unlimited numbers of channels of video, telephony, switched data, wide area networking, business data links, etc. These networks will carry services (sometimes called video dial tone) including broadcast video, time-shifted TV, pay-per-view TV, multimedia services, premium channels, video on demand, home shopping, tele-education, interactive games, etc. This interconnected high-performance network is often referred to as the Information Superhighway, or the National Information Infrastructure (NII).

These NII component networks will be coax or hybrid coax and fiber-optic links, creating a single network delivering a multitude of services with integrated diagnostic capability. The network elements will operate in the 5-MHz to 1000-MHz range, employing ATM digital packetized communications and switching protocols and making access possible to a wide variety of service providers. Multiple elements will be interconnected and will create additional capacity. The Information Superhighway will open up what has been estimated by some to be a trillion dollar delivery system for offering goods and services. Interactive TV (ITV) will be a highly efficient merchandising and distribution system.

The emerging technologies making this possible are advanced microprocessors, satellites, microwave, fiber optics, new switching technologies, image compression, video servers, and high-capacity video-friendly disk storage systems.

Both the telephone companies (TELCOs) and the cable TV (CATV) companies are pursuing the coming generation of ITV. The TELCOs have traditionally supported interactivity, and the CATV companies have supported multiple television broadcasts. This new technology has opened up the doors to both the TELCOs and the CATV companies to furnish ITV. The TELCOs now have a new technology called Asymmetric Data Subscriber Loop (ADSL), permitting the transmission of 2 to 6 MHz of bidirectional data bandwidth to most subscribers on existing twisted pair cables. Theoretically, the TELCOs could provide different programming on each pair of twisted wires. Every subscriber could have a different program, but having sufficient video server capacity at the central office would not be cost-effective. The CATV companies have gigantic one-way bandwidth with little experience in bidirectional communications and switching. The TELCOs have little experience in broadcast technology, but TELCO and CATV companies are definitely on this convergence path, as partners or adversaries.

Cable will be a vehicle for serving ITV. The cable TV (CATV) industry already serves 58 million households with a flexible, high-capacity fiber-coaxial network capable of delivering a variety of video, computer, and voice communications data. The

TELCOs provide wired services to over 94 million households. The advent of ADSL communications technology facilitates the transmission of video over several miles of conventional 22-gauge wire, along with plain old telephone services and a separate return medium-speed digital data path, creating possibilities for an alternative technology to deliver images, video, and data to households and businesses.

Since the advent of the microprocessor in 1968, computer system costs have been spiraling downwards, and computer intelligence has migrated into storage subsystems, imaging, communications, and network control. Subsystems permitting complex system advances were previously not possible. High-capacity advanced random access storage subsystems now permit access to multiple gigabytes of data, and permit storing entire films compressed onto a single computer disk. Video servers consisting of hundreds of these high-capacity disks permit selection of video programming from hundreds of titles. Because of the interactivity of this storage medium, interactive television now can become a reality. This concept extends from CATV and TELCO systems, to hotel pay-per-view systems, to video on demand for aircraft entertainment systems, which will provide individual programming and individualized services to airline passengers.

But the cost of the system components dictates whether the system can be economically viable. Enhancements to storage technology are evidenced by video-friendly disks (sometimes called audio-video disks) that permit the smooth flow of video from storage to the network without extensive and expensive buffering. Video server technology is moving from an expensive brute-force approach to an architecturally streamlined approach, taking advantage of video and data differences and exploiting them to reduce video server cost.

From a communications point of view, image compression is a must; it permits 4 to 6 digital channels to exist where only one analog channel previously resided. From

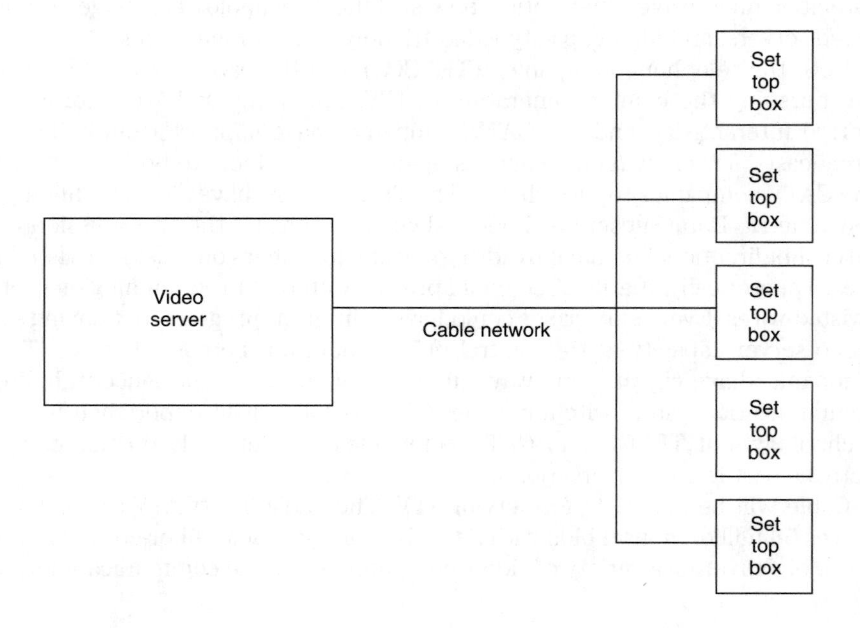

a storage point of view, uncompressed digitized images require hundreds of times more data than compressed data. From a networking point of view, fewer data packets are required for transmission if the images and data are compressed, thereby permitting reasonable cable bandwidth and resulting feasibility.

Convergence of technologies in the cable system

Cable television companies have built a very efficient delivery mechanism, with a wide variety of video entertainment options. As technology continues to evolve, the integration of previously separate industries has not only enabled cable to diversify, but has also enabled entirely new types of interactive multimedia services to be contemplated and new business arrangements to be formed in order to deliver them. Some examples follow:

Example #1: The Time-Warner Full Service Network uses as the core building block of its project the existing cable television system. That system is being upgraded according to an emerging cable industry consensus, calling for an evolutionary upgrade of plant by deploying fiber optics in a cost-effective way out to serving areas with as few as 500 homes. This upgrade, which by itself improves picture quality, increases system reliability, reduces maintenance costs, and sets the stage for other technologies to be connected to it or overlaid on it.

Example #2: A video server is being provided by Silicon Graphics, a company with experience in high-end computer workstations. The server will hold a digital library of hundreds of movies and other multimedia products that consumers will be able to choose from their homes. Videos will be routed from the server to individual homes by an asynchronous transfer mode (ATM) switch. ATM switching is a recently developed protocol for switching large amounts of digital data at high speeds, and American Telephone and Telegraph (AT&T) will be providing the switch in Orlando, Florida. The consumer will also see advanced technology deployed in the home. Hewlett Packard, a leading computer technology company, will be providing printers in consumer homes, allowing consumers the convenience of capturing information that appears in text, video, or multimedia forms on their TV screens.

Example #3: In some smaller cable systems, it may not be cost-effective to deploy all of this advanced technology at once. With fiber upgrades, however, many small systems may be able to interconnect through a design known as a regional hub and share the costs of advanced technology. The regional hub design calls for linking head ends through a fiber ring, such as those provided by businesses known as competitive access providers. (Cable companies have invested in such companies, and the largest such company, Teleport Communications Group, is owned by a consortium of 5 cable operators.) The head ends are linked in this fashion into a regional hub that allows advanced technology to be located in the hub and still be available in a cost-effective manner for smaller head end service areas.

Example #4: In addition to telecommunications services, the wide array of high quality programming provided by cable programming networks is being enhanced by the application of various kinds of processing and interactive technologies that are becoming available. Viacom, for instance, will be testing interactive versions of networks like MTV and Nickelodeon in their Castro Valley, California research test bed cable system.

Example #5: The Discovery Network has developed Your Choice TV, an advanced program selection and navigation system that will allow customers to find and select a variety of programming, including "best of" television programs that they might have missed in their regularly scheduled network appearance.

Example #6: Many cable programming networks now make detailed programming information and schedules available on Prodigy, the computer information service.

The obvious synergies of combining CATV/TELCO ITV systems with computer processing power and the diverse networking abilities of telecommunications companies are also contributing to this convergence.

Computer companies

The following is a partial list of computer companies that are offering products specifically designed to enhance the ability of cable systems to deliver a multiplicity of interactive multimedia products and services to their subscribers. They include:

SEGA. Sega, the video game company, is starting the Sega Channel, a cable channel in conjunction with TCI and Time-Warner, creating a cable channel that is actually devoted to video games that can be played at home without requiring physical possession of a game cartridge.

Intel. Intel has developed, in conjunction with General Instrument, a cable modem that allows data to be received and transmitted over cable systems.

Digital Equipment Corporation (DEC). DEC has developed a "Collaborative Work Environment" that can be offered over a cable system through the use of DEC's Digital Channel technology, which turns part of the system into an Ethernet wide area network. This technology allows multiple participants to work together to develop computer-assisted design applications or distributed professional work group computing (which allows teams of people to organize and update information in the same way, no matter where they're located).

Hewlett Packard (HP). HP has three sets of products that they are developing for use in cable systems. HP recently received an order from Tele-Communications, Inc.(TCI) for 100,000 set top converter boxes; these boxes will include advanced computer processing capabilities, permitting customers to use on-screen menus to find and select video or other services. HP also offers video server technology, computers that can be used to store movies and other programs in digital form for video on demand. The server may also include ad insertion, store-and-forward, editing, and other capabilities. Finally, HP has developed a video printer and print manager that will first be rolled out in Time Warner's Orlando system. The printer will be used for capturing and printing out screen displays as well as printing out coupons at the customer's request.

Eastman Kodak. Kodak's Picture Exchange architecture and printers allow image search, retrieval, and printing over cable systems. Customers can view image libraries (which may be connected from a central location to cable systems by high speed transmission lines) and select particular images for printing at the home or at centers where they can be picked up.

CompuServe/Prodigy/America Online. The computer information services America Online, CompuServe, and Prodigy are planning to offer service over cable systems, allowing huge increases in speed and quality of the service. General Instrument, Zenith, and Intel are providing equipment as part of these trials, and cable operators participating include Viacom, Comcast, Cox, and Continental.

UNISYS. UNISYS provides desktop video conferencing systems allowing individual users to view a common video or data display while in different locations. The users can be connected via standard Ethernet connections over a cable system.

IBM. IBM has developed a multimedia server that allows storage and delivery of various video-on-demand services. Connection to services such as ICTV's Interactive Service Exchange allows customers to instantaneously switch from viewing a motion picture to using an interactive service.

C-Cube. C-Cube has demonstrated its Motion Picture Engineering Group (MPEG) encoder/decoder products, which allow transmission of digitally compressed video to increase channel efficiency of existing cable networks, as well as to make more efficient use of fiber and other digital facilities.

Zenith. Zenith provides products that allow Ethernet services to be delivered over cable television systems. The benefits include connecting institutional networks of schools and government agencies, and allowing individuals to use the cable system for high-speed on-line connection to information services without tying up their phone line. Individuals can also access their office networks through work at home applications.

Micropolis. Micropolis has developed video-friendly disk drives for use in video servers. Unlike conventional computer disk drives, the Micropolis disk drives provide a smooth flow of multiple thread audio and video data.

Innovative telecommunications services

Telecommunications services ranging from plain old voice telephone service, to access services connecting customers to their long-distance carrier of choice, to data transmission, and all the way to video telephony are services that are now being deployed in the context of cable television architecture. These services also bring into the cable family a range of new technologies and services provided by companies that in many cases were not previously involved with the cable industry.

NEC. NEC's Asynchronous Transfer Mode (ATM) switch provides video and broadband switching capability that can process voice, video, and data, and can route different types of information to different customers at the customer's request.

Northern Telecom. Northern Telecom designs and builds high-speed (2.5 gigabit/second) fiber rings such as those used in providing competitive access services, which can be used for interconnecting cable head ends. Northern Telecom also provides ATM switching technology as well as network management for all services provided by its fiber rings.

DSC Communications Corporation. DSC provides network support for systems based on the SONET standard, allowing transport of existing and emerging applications in the business community. DSC network support can connect a Competitive Access Provider (CAP) network to a cable company transmission network that is connected to a local business site, allowing that business to take advantage of new or more efficient services without being directly connected to that CAP.

Scientific Atlanta. Scientific Atlanta, a traditional supplier of cable system equipment, has announced the development of a product that will allow consumers to make telephone calls over their cable television systems.

About the book

This book will visit the issues of technologies, facilities, functions, user expectations and costs. Conclusions will be drawn. This book will describe all the functional system building blocks as diagrammed below, including the interactive return paths which are not shown.

The programming will come from a source such as videotape, film, or live from a camera. The signal digitization, preprocessing, encoding, storage, transmission, decoding, postprocessing, and ultimate viewer display technologies will be discussed and compared. The final section of the book is a glossary of terms used both in the book and in the industry.

The reader will observe a slight redundancy between certain chapters. This was done to permit the chapters to be stand-alone units as much as possible, and to permit the reader to begin the book somewhere in the middle and randomly access the rest of the chapters, if desired. For example, chapter 8 on storage systems and video servers discusses the interface to the communications channel, as does chapter 11, the "ITV Set Top Box," and a separate chapter exists to deal with the communications channel. Other similar elements of redundancy exist between certain chapters to assist the reader in the expedient pursuit of his or her desired ITV reading interests, and to provide a continuity between chapters.

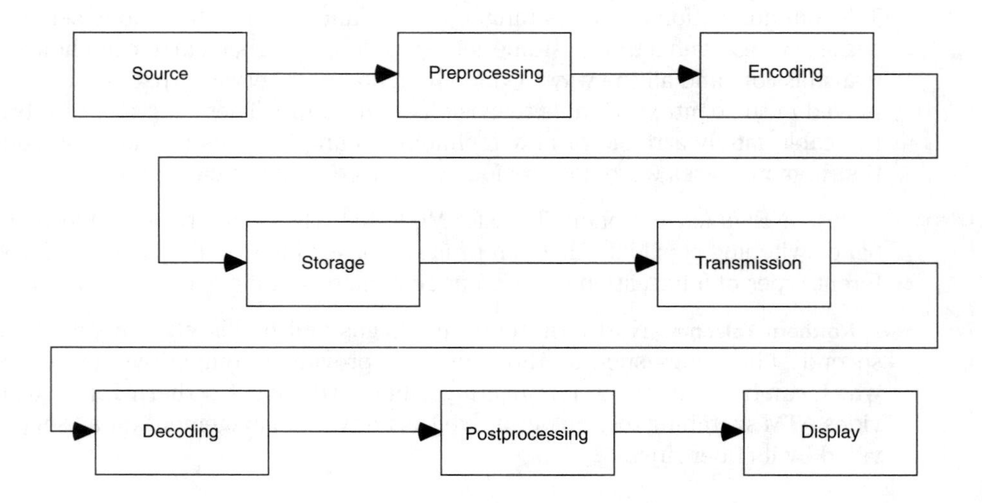

Chapter

1

The Merging of Computers, Imaging, Communications, and TV (Convergence)

This book describes interactive television, advanced multimedia, and the convergence of television and computers. It discusses the requirements for interactive television (ITV) systems and the requirements for advanced multimedia systems. It will discuss the shortcomings of today's computer-based multimedia and how those inadequacies can be corrected. It will introduce the concept of advanced image storage systems (video-friendly disk drives and video servers), compression requirements, communications, switching, and set top boxes. Philosophy and standards will be discussed, and the key players and their technologies will be examined. Whether it is a cable television (CATV) operator or telephone company (TELCO) providing interactive television to the homeowner, or whether it is a business with its own advanced multimedia system comprised of its own server, network, and computers, this book will show how these are essentially the same technology. This book will culminate with some serious speculation on the future of this convergence, but first, if we look at the short history of media broadcasting, it can provide us some insight on where convergence might be headed.

Electronics provided the communications possibilities and the evolution of communications began its upwards spiral. Communications started with pulsed or switched coded signaling, followed by the application of acoustic modulation to a radio frequency carrier, and then to the application of image transmission. Today, computers are well integrated with the communications processes of encoding, enhancing, controlling, compressing, and so forth.

Comparing the single transistor on a chip of yesteryear with today's millions of transistors on a chip, and the expectations of billions of transistors per chip in the early part of the next century, system growth in capacity and complexity with continuing cost erosion is expected.

In 1885 Guglielmo Marconi invented radio telegraphy based on the findings of Michael Faraday that a spark created via an induction coil could create another spark across a similar spark induction apparatus across a room. Even though there was no physical connection, the two events were correlated. This was the birth of intentionally made man-made electromagnetic fields. It wasn't much later that Marconi discovered that by turning these fields on and off, one could communicate across long distances. These were radio signals, the same ones modulated today with video, data, voice, music, telemetry, and so on.

The vacuum tube and transistor, devices giving us better control over the flow of electrons, permitted the creation of more efficient radio equipment. These same devices have permitted the logic implementation in computers and control systems. Vacuum tubes, cumbersome, hot, and sometimes consuming 3 or 4 cubic inches, have given way to the transistor and integrated circuit. The integrated circuit can provide several million transistors packaged into a tiny fraction of a cubic inch. Voltage and current-controlling devices have shrunk by about 10 million to 1 in 70 years, functional complexity has increased accordingly, and price has followed in an inversely proportional manner. These are the building blocks of command, control, and communications systems. These devices are permitting us to enter the era of interactive television and advanced multimedia.

TABLE 1.1 A Summary of Significant Communications History Events

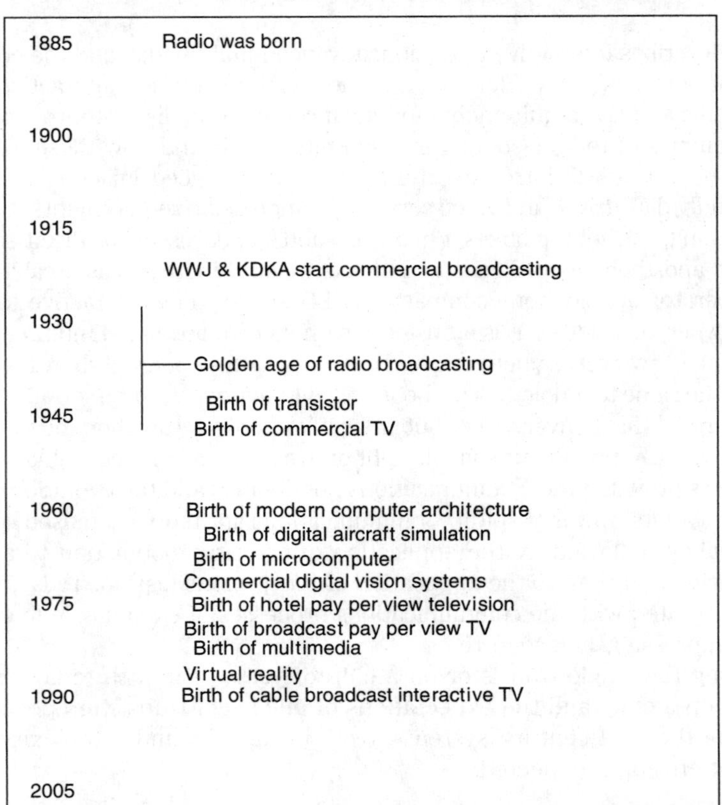

1885	Radio was born
1900	
1915	
	WWJ & KDKA start commercial broadcasting
1930	
	Golden age of radio broadcasting
	Birth of transistor
1945	Birth of commercial TV
1960	Birth of modern computer architecture
	Birth of digital aircraft simulation
	Birth of microcomputer
	Commercial digital vision systems
1975	Birth of hotel pay per view television
	Birth of broadcast pay per view TV
	Birth of multimedia
	Virtual reality
1990	Birth of cable broadcast interactive TV
2005	

In 1962, computers and TV projection systems were first combined to create simulated environments for training aircraft pilots and testing new aircraft even before the planes were constructed. These simulations created video projections on a screen in front of a realistic aircraft cockpit mockup. As crude as they were, with a little imagination the pilot could believe he or she was flying the airplane and that the airplane was responding to his or her commands. To make the simulation more realistic, the entire cockpit could be moved under computer control to simulate roll, pitch, and yaw. This was not only the birth of Interactive TV (ITV); this was the birth of virtual reality and advanced multimedia. Such a simulation required millions of dollars worth of computers, TV projection equipment, and hydraulics, coupled to a special environment in the cockpit. Today, excluding the physical positioning equipment, comparable experiences can be produced thousands of times less expensively.

ITV can be defined as TV that is controlled by the viewer, whether this implies interprogram or intraprogram decision-making capabilities. The difference is only the degree of interactivity. Interprogram ITV could be defined as the selection of new video on demand (VOD) programming, such as selecting a new movie when the viewer wanted it, with minimum waiting time (from seconds to minutes), but with intervening information that made the viewer's requests appear immediately responsive. Intraprogram decision making would occur frequently in educational ITV applications, but not so frequently in filmed entertainment applications due to the high cost of producing interactive film presentations with alternative scene-to-scene jumping possibilities. Interactive games would also exploit the ITV interactivity.

Advanced multimedia can be defined as the highest quality audio and visual presentations with very frequent participant interactivity. Thus, advanced multimedia can be thought of as ITV with a higher frequency of participant interaction. The difference between ITV and advanced multimedia is essentially the degree of interaction. Generally, an advanced multimedia application will be a single-user application and an ITV application will be a multiuser application. These concepts overlap and converge.

The similarities between interactive TV and advanced multimedia are so significant that this book spends a great deal of time revealing those similarities. Figure 1.1 illustrates an advanced multimedia system on a local area network (LAN), while Figure 1.2 illustrates two multimedia LANs being interconnected through bridges or routers (brouters). The brouters keep the traffic destined for a specific LAN contained, while permitting only LAN-to-LAN traffic to pass through. If there were not any internetwork communications, the brouter would be unnecessary.

Advanced multimedia networks provide interactive, full-performance, smooth-motion video to be displayed on each workstation just as if it were a TV set, and while these workstations could display favorite movies, it is more likely that they would be displaying training information, tutorials, research data, and other business-related information. For example, Figure 1.1 could be a system entirely resident at an automobile dealership, where some of the workstations are in the showroom for customers to interact with in the car selection process, while other workstations could be in the service department being used by mechanics to investigate correct procedures for repairing a specific automotive function. At the same time, a salesperson could be investigating the merits of an automobile versus the competitors' automo-

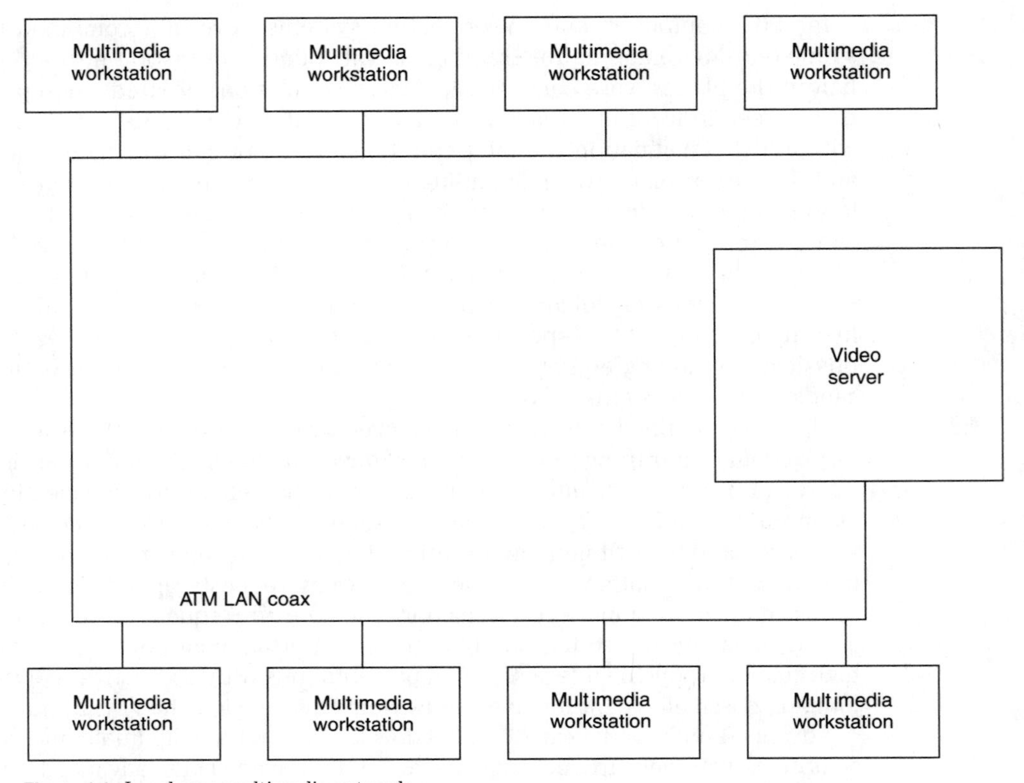

Figure 1.1 Local area multimedia network.

biles. This system could be a small self-contained video server LAN, or it could be a group of automotive dealers' systems interconnected so that information could be shared.

Advanced multimedia systems can be grown to include virtually any number of workstations and any number of video servers. Their usefulness is limited only by the total traffic on the network, and hence by the network and video server availability to the user at his or her workstation. The systems shown as Figures 1.1 and 1.2 are generally delivering unique streams of data to each workstation, unlike a CATV system, which can deliver one set of data to many workstations/terminals/TV sets. While it would be possible to have advanced multimedia workstations share data streams, it is probably not too likely.

Figure 1.3 represents an ITV system with a video server connected. Each TV can interact with the network to permit the viewer's selection of desired programming. Once the TV programming selection request process has been made, the viewer is connected to his or her program. If a 90-minute program is rebroadcast in a time-overlapping manner every 6 minutes, then 15 copies of the program will need to be broadcast with a 6-minute delay between broadcasts or program threads.

Variations on the channel capacity of different CATV systems can be speculated, assuming: (1) digital TV program compression, (2) 200 MHz to 500 MHz of digital

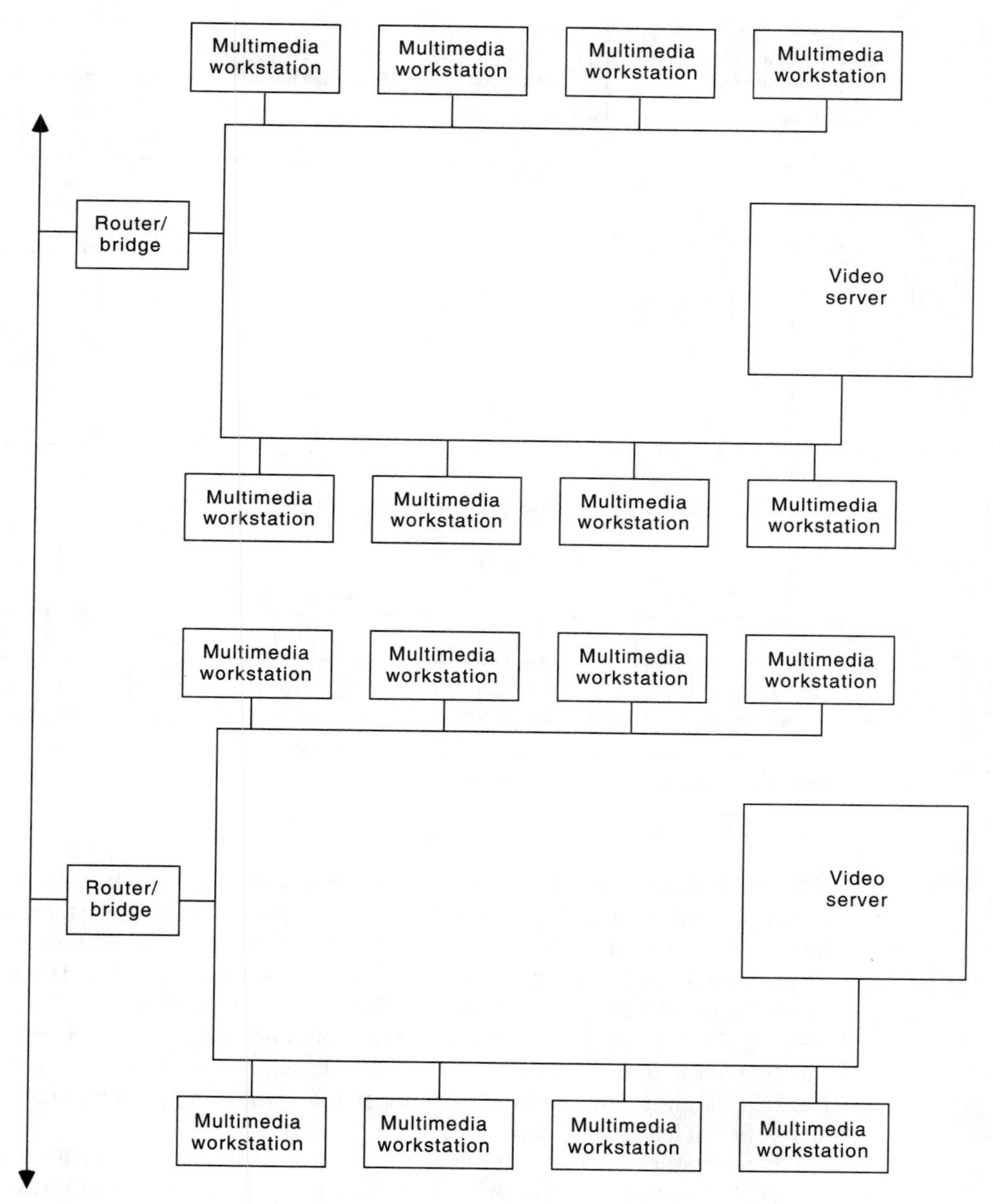

Figure 1.2 Wide(r) area multimedia network.

transmission capacity; this book will show how a range of 40 to 500 programming channels or threads can be made available on a CATV system without affecting existing analog television channel transmission. Naturally, a multicoax CATV system can produce multiples of the above, and an all-digital system (one where all programming is compressed) could support even more threads.

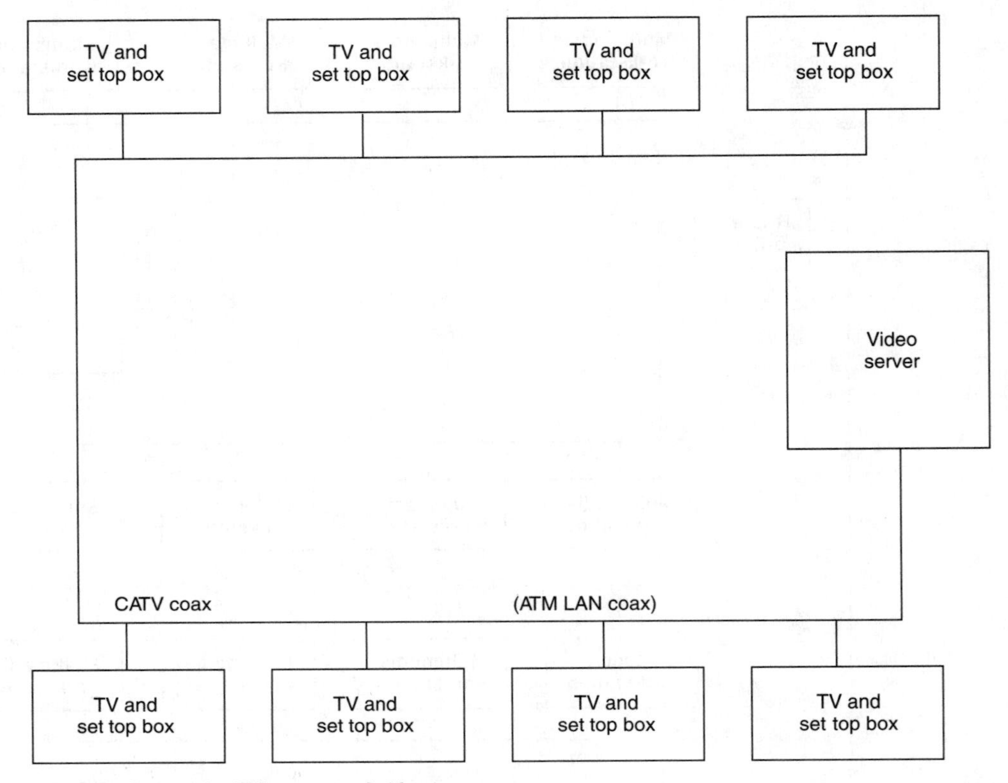

Figure 1.3 Interactive TV system and video server.

There will come the time when multiple fiber-optic cables intertwine the cities we live in, and bandwidth in excess of the requirements for ITV will exist. On such a system, it will be natural and convenient to mix ITV and advanced multimedia programming. Such a system might appear as in Figure 1.4. Figure 1.5 depicts a generalized view of a mixed advanced multimedia / ITV system. The limitations on such a system are only the available bandwidth and utilization. Because the network bandwidth is a function of the number of advanced multimedia and ITV devices and their usage, it is possible to determine the number of users that can be supported under various assumptions of loading.

When clusters of users and servers are broken down into smaller units, more bandwidth is available to each user. Breaking down very large systems into smaller micro-clusters is a method to resolve limited bandwidth. CATV coaxial cable and repeater systems have a limited bandwidth interval ranging from 50 MHz to a maximum of 1 GHz, and thus can support a maximum of 1000 VHS tape deck-quality resolution programs to 100 very high quality programs, or 150 conventional analog TV programs. Added capacity requires optical fiber cables.

The secret to nodal clustering is to restrict information flow between an internal server and its user to that specific cluster, and not to permit this otherwise unnecessary information to pass externally to another cluster unless that information is required by another user, outside that specific cluster.

Significant cluster-to-cluster communications will require the higher bandwidth of optical fiber cable and gateways (or switches) to provide intercluster communications. This is discussed further in chapter 10.

Light Highways (Data Superhighways and the National Information Infrastructure)

One can imagine a nationwide or worldwide network of clusters connected via a fiber-optic light highway called, in the U.S., the "National Information Infrastructure (NII)." This worldwide infrastructure would support the viewing of events over long distances, exchanging pertinent government, business, personal data, or medical information, conducting telephone calls, performing electronic publishing through remote multimedia, etc.

Video servers

This chapter so far has only alluded to the role of the video server. The video server is a large array of video-friendly disks that stores gigabytes of media programs. Typ-

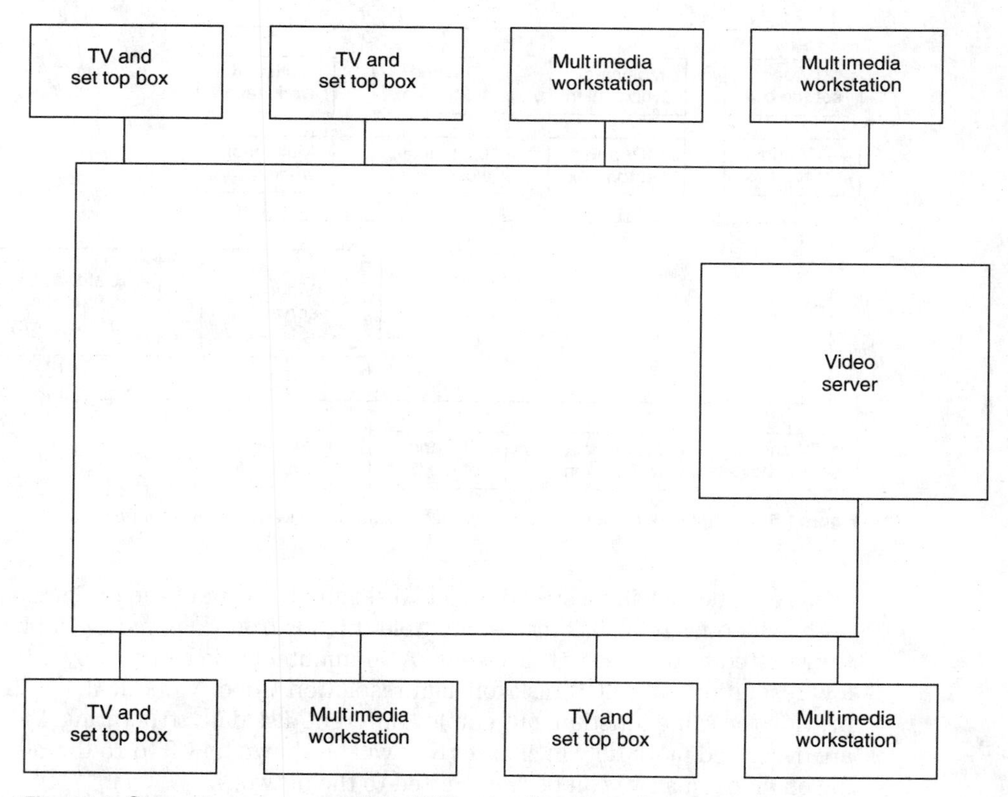

Figure 1.4 Composite system consisting of ITV and multimedia applications.

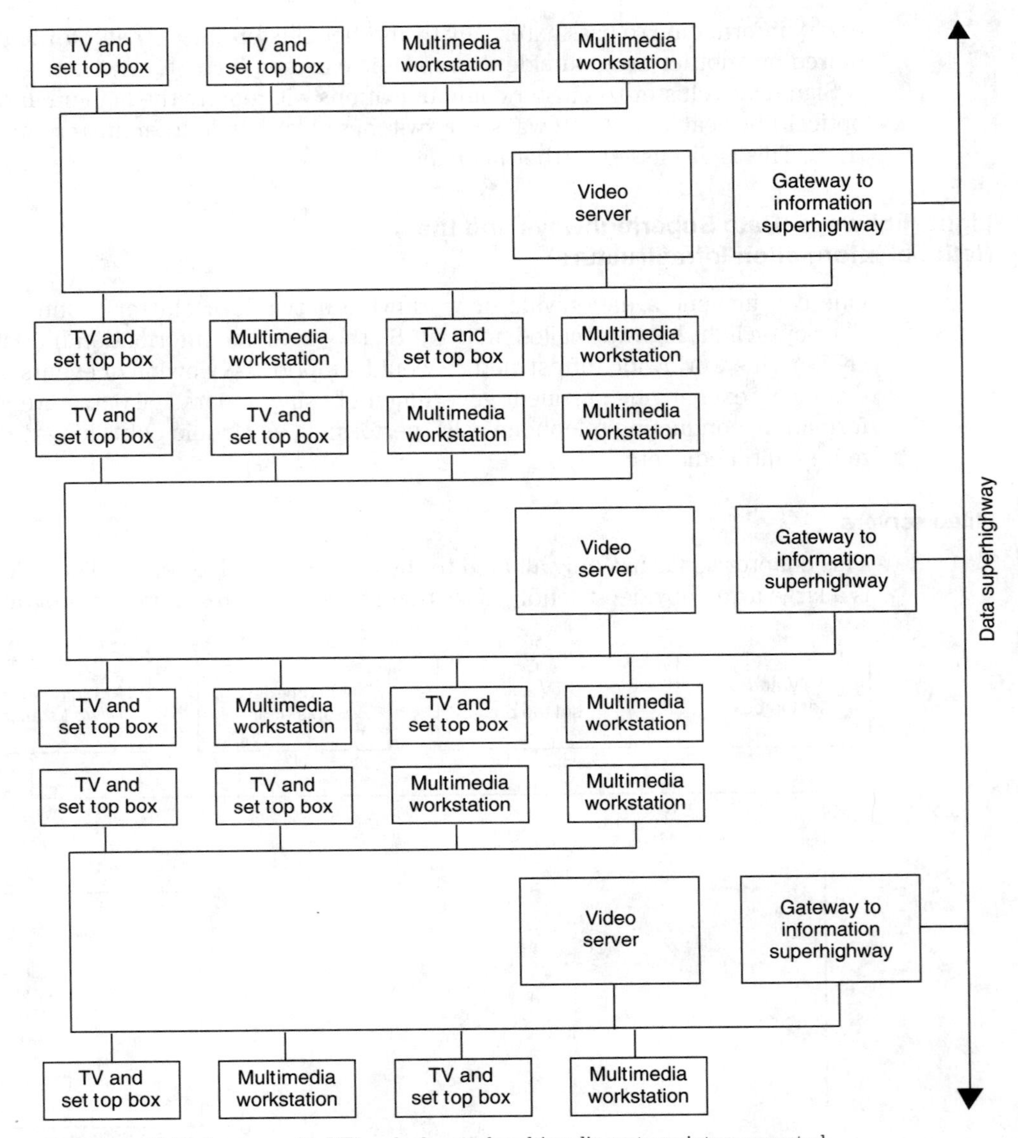

Figure 1.5 Multiple composite ITV and advanced multimedia systems interconnected.

ically, one movie will be stored on one disk, and because of the design and capacity of the video-friendly disk drives, multiple threads from each movie can be read and transmitted to the rest of the system. A 90-minute movie can occupy a 1GB disk for low resolution, or a 6GB disk for high-resolution video. Video RAID systems could provide the same function, but employ different disk data structuring. Video servers are described more fully in chapter 8. It will be shown how 3 to 15 threads (depending on video quality) can be transmitted to the network.

Computer terminals and TV set top boxes

As we have seen in this chapter, the ATM network does not care whether it is talking to a multimedia computer or a TV set top box. Both units are involved in selecting and controlling media. Both units handle the ATM protocol, signal demodulation, and decoding. Both units can share a 5-part VLSI chip set.

An Overview of CATV Broadband Integrated Services Network Development

The cable television industry in the U.S. has about 59 million subscribers and passes by about 93 million homes. There are more than 11,000 cable head ends, and the cable TV industry has installed more than 1,000,000 network miles, with fiber-optic cabling in the U.S. growing from 24,000 miles in 1992 to more than 41,000 miles in 1993. Construction spending by the cable industry in 1993 was estimated to be about $1.9 billion, of which $1.2 billion was for rebuilds and upgrades. Cable industry revenue from subscriber services in 1993 was estimated to be more than $22 billion, corresponding to an average subscriber rate of about $31. These figures are expected to grow as the cable industry moves into new interactive digital services and telecommunications. The convergence of cable television systems, telecommunications systems, and computer network systems will drive the future information age, and integrated services network architectures will provide the key enabling technology for the development of this new business sector.

The cable industry's development and deployment of the broadband information superhighway requires architectures for a network infrastructure that integrate the numerous applications that will be offered. Applications will cover a wide range of programs and services. Entertainment video delivery will evolve from the current core services of the cable industry to enhanced offerings like interactive shopping, near video on demand, and true video on demand functions. Telecommunications services will evolve from current voice telephony and data transport to include interactive multimedia applications, information access services, distance learning, remote medical diagnostics and evaluations, computer-supported collaborative work, and more.

In addition to the complexity and diversity of the applications, development and deployment of a broadband information infrastructure will combine a number of different networks that will have to work in a coherent manner. Not only will users be connected to different regional networks, but the sources of information will also belong to different enterprises and may be located in remote networks. It is important, therefore, to realize from the start that the two most important goals of the architecture for the broadband information superhighway are integration and interoperability.

Achieving these goals is a mind-boggling endeavor. Nonetheless, it is a critical goal that goes beyond just elegance of the technical solutions. First, the investments that will be required to realize these networked broadband applications will be quite large. It will be difficult to economically justify the widespread deployment of disjointed networks. The value of communication services grows exponentially with the number of interconnected users, and future broadband applications will have to reach globally to large numbers of customers in order to be commercially viable.

Second, the multiplicity of broadband applications that one can envision in the future will not emerge from a centralized planning and deployment process, but rather from an information marketplace where independent application developers will compete for the customer's attention. The existence of integrated and interoperable networks will stimulate the rapid development of new applications, and hence speed the introduction of new services by providing added value to users.

The cable industry's broadband integrated services network architecture should be based on a hierarchical deployment of network elements interconnected by broadband fiber-optic and coaxial cable links. Following is a view of this architecture: Starting at the home, a coaxial cable tree-and-branch plant provides broadband two-way access to the network. The local access coaxial cable plant is aggregated at a fiber node, which marks the point in the network where fiber optics becomes the broadband transmission medium (see Figure 3.1, p. 51). Current expectations are that approximately 500 homes will be passed by the coaxial cable plant for every fiber node, with variations (from as low as 100 to as many as 3000) that depend on the density of homes and the degree of penetration of broadband services. The multiple links from the fiber nodes reach the head end, which is where existing cable systems have installed equipment for origination, reception, and distribution of television programming and other services. The head ends are in buildings that provide weather protection and powering, and hence represent the first natural place in the network where complex switching and processing equipment can be conveniently located. Traffic from multiple head ends (see Figure 4.8, p. 69) can be further routed over fiber optics to regional hub nodes deeper in the network, where capital-intensive functions can be shared in an efficient way.

In order to achieve the integration and interoperability of the different services, the first basic issue to consider is bandwidth allocation. Today's cable networks use the coaxial cable plant as an RF medium, and generally allocate spectrum as follows:

1. 50–450 MHz
2. 50–550 MHz
3. 50–750 MHz
4. 50–1000 MHz

for the downstream traffic (toward the users); and

1. 5–30 MHz
2. 5–35 MHz
3. 5–42 MHz

for the upstream traffic (low-split approach, from the users). See chapter 10, ITV communications.

Interoperability is expected to be facilitated by the use of industry standard technologies, in particular synchronous optical network (SONET) networking elements and asynchronous transfer mode (ATM) multiplexing and switching. It is possible that ATM will eventually become the prevalent solution all the way to the user interface at the home, but alternatives are being explored for the early deployment of broadband services in a cost-effective manner. Other mechanisms for internetworking will have to be used until competitively priced ATM products, including interfaces, multiplexers, and switches, become widely available over a range of speeds from a few MB/s to GB/s. Issues that still need to be resolved in ATM are numerous: access and congestion control, quality of service definitions, choice of adaptation layer standards, transport of compressed packet video (MPEG in particular), end-to-end latency with regard to isochronous services (voice communications in particular), advanced signaling protocols and connection control systems, and management information models and interfaces. The final evolution to an integrated services broadband network will require consideration of many additional interfaces and protocols associated with the many services and network interconnection points that will be supported. Signaling protocols, directory access services, and application development environments are only some of the issues to be resolved.

Integrated network management, operation support, and business support systems are important components of the broadband architecture. The use of open and modular platforms and protocols for the development of support systems is a crucial factor in controlling the complexity that the multitude of network components and applications will create for the network operators. The evolution should be toward the use of distributed computing environments, coupled with modern software technologies such as managed object models, client-server architectures, and contract trading capabilities.

Chapter

2

Philosophy, Possibilities, Architecture, Concepts, Expectations, and Standards

Interactive television, advanced multimedia, and virtual reality are on a convergence path. What makes the evolved version of each of these different is the application. When a suburban CATV subscriber selects a movie, orders some food, and orders some specialized information from his ITV system, it is referred to as interactive television. When a physician in the hospital uses his or her personal computer attached to the very same network to watch instructional sequences of specific operational procedures, with permitted branching at certain sequences of the operation, it is called advanced multimedia. Virtual reality can be consumer- and/or business-based, usually providing higher quality human perceptions. Interactive educational systems can be any of the preceding. The basic difference between the various applications is the degree of interactivity and perception quality. All can use the same network facilities, the same video servers, and much of the same circuitry at the user end.

Video on Demand: An Opening Scenario

To create a human factors experiment to test our conceived ITV system, we will create a mythical ITV consumer sitting at his or her interactive TV set. Assume the consumer is pondering what would be entertaining today. He or she picks up the very simple one-button remote control. As the user picks it up, a menu appears across the top of the TV screen with a series of small rectangles; each of these blocks has information in it. They look like the screen below.

Our subject decides to watch a movie, so he or she points the one-button remote control (with integrated laser pointer) and points to "movies," and the menu extends beneath movies into 5 groups. The available choices could be Romance, Action-

Figure 2.1 ITV menu system on TV set.

Thrillers, Classics, Fantasy, and Comedies. The user chooses "Action-Thrillers," and near the top of the list could be "Cliffhanger" with Stallone. This is the movie the viewer wants to watch. He or she requests the movie and, just as he does, five tempting pizzas appear on the screen: pepperoni, vegetarian, Sicilian, cheese only, and a combination pizza. The user is instructed to point and click the one-button remote on the pizza of choice if, of course, he or she wants a pizza. He or she is instructed that the pizza will be delivered to the door within 30 minutes of the button pressing, or it will be free. The user doesn't have to tell anyone where to deliver the pizza because *the system knows who and where he or she is*. Then comes the beverage selection. The user selects a beverage and gets ready for the movie. The complete food order on the screen appears for validation purposes. Also, a few excerpts from the movie appear, in order to validate that this is really the movie he or she wants to see and giving the viewer an opportunity to change his or her mind. The user is then given pricing information, and the movie is ready to start.

About two minutes have lapsed from the beginning of this transaction to the beginning of the movie. The user has been entertained and his or her needs have been catered to during this entire interval. The movie is starting, the user is replenished and invigorated, and he or she is ready for serious movie watching. Within thirty minutes into the movie, the doorbell rings. It's the pizza delivery person. The user gets up from the TV and does not need to get any money; the system has already charged the pizza to an ITV or credit account. The viewer goes to the door, tips the pizza deliverer, gets the pizza, goes to the kitchen for some eating utensils, and returns to the movie.

However, the viewer has lost three minutes of the best part of the movie. He or she picks up the remote control and immediately sees fifteen small screens on the big screen, each showing the same movie, but with different starting intervals. The user selects the interval subsequent to the one he or she had been viewing (the movie, delayed by three minutes) because he or she recognized it in the small screen as the part of the movie that was shown just before the pizza arrived. The user resumes watching the movie from where he or she had previously been interrupted.

Our subject settles back and starts to eat pizza and, about another twelve minutes into the movie, there is a scene that would certainly nauseate any food-consuming connoisseur. He or she picks up the remote control, again the small screens appear on the large screen, and the viewer clicks on the small screen that has just completed this gruesome sequence. The user bypasses the grisly sequence faster and easier than he or she could fast forward a VCR, because he or she has essentially random access to any scene, on a scene-by-scene basis. The user continues to watch to the completion of the movie.

Because the cable TV system employs an open architecture, some of the movies will be encoded as MPEG 2, or as wavelets, or using spatial correlation technology. This permits the movie studio to select the best encoding for the specific movie, because different compression algorithms create different artifacts as the compression requirements become more demanding. The action scenes with lots of fast motion are the scenes most likely to suffer. It may be that no significant difference can be perceived by viewing the different algorithms. However, the studio executives may have their preference and, of course, this is their film, so the selection of compression methodology is their choice. The studios perceive compressing video, digitizing it, and placing it on some information-bearing substance as their domain, just as is putting movies on VHS tapes. Furthermore, the studio executives realize that the various standards committees find it difficult to agree on specific compression standards, and that ownership of certain compression technologies may be questionable at best. So the studios create their own compression algorithms, digitize the film, and include decompressing information in the form of firmware for our viewer's TV set top box to decipher and execute.

There is more. The studios, concerned with theft of their product, have concluded it is necessary to encipher the program before they send it to the viewers so that the set top boxes will have to decipher the signals before they can be decompressed and viewed. Our viewer's box has been previously given a special key so that it can unlock this movie he is watching.

This small set top box, which our CATV viewer purchased for $150, is a relatively busy little box. After it transmits a movie request to the cable TV company, then finds the digitally encoded signal destined to the user with his or her movie, decodes the ATM protocol, checks for transmission errors, deciphers the signal, decompresses both video and audio, and creates the signal formats for the TV and hi-fi system, the user discovers that the box is performing a multitude of other functions! The user's set top box, connected via low-power wireless technology, has also been talking to the water meter, electric meter, gas meter, home security system, and environmental sensors and control systems.

Someone is trying to reach the user on the phone. The house telephone wiring is connected to the RJ-11 connector on the back of the set top box. Next to the RJ-11 connector is a 244KB RS422 connector, which is linked to the user's Power PC. The user is connected to wide area network (via the national information infrastructure, NII) with performance far greater than any local ETHERNET employed a couple of years ago.

The back of the TV set top box looks like as shown at bottom of page:

Of course, this is 1997, and the user just upgraded their box to this year's premium model. The 1995 model didn't have the RJ11 and RS422 connections.

Now our hypothetical hero(ine) plays games with distant players, has almost instantaneous response to computers in other states or countries, and can shop in any major store on the NII...

...End of scenario.

By going through the above scenario, it is possible to create a list of system functions and define where those functions reside. Table 3.1 is such a list. An asterisk in the columns represents where some or all of the functional contribution exists.

A discussion of each of these individual functions follows.

Remote control. The remote control should be as simple as possible to operate. Unlike VCRs, whose operation is difficult at best, the remote control should be simple, intuitive, and practical. A single-button remote that the user merely points to his or her desired selection on the screen meets these criteria. The functions that can be supported by the remote device should include:

- Program selection
- Viewing time selection
- System responses (diagnostics, status, error processing, control, etc.)
- Order placing
- Menu navigation
- Home shopping

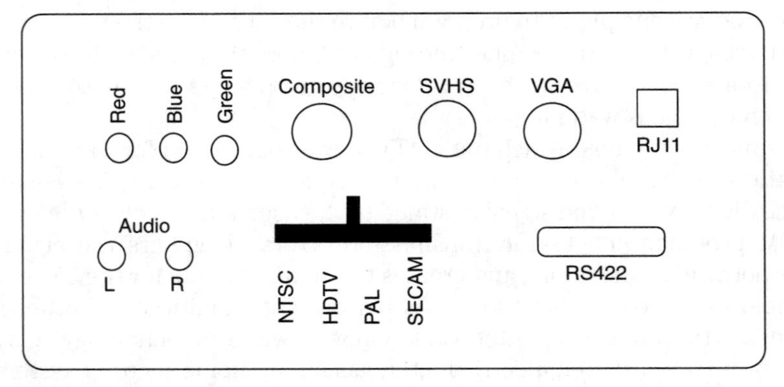

Figure 2.2 Rear view of TV set top box.

Function	System	Set Top Box	Head End	Communications
Remote Control		*		
Movie Selection	*	*	*	*
Navigation	*	*	*	*
Menus	*	*	*	*
Interactivity - Shopping	*	*	*	*
Interactivity - Games	*	*	*	*
Addressing - Real	*	*	*	
Addressing - Virtual	*	*	*	
Object Selection from menu		*	*	
Windowing	*	*	*	*
Program Interruption		*		*
Scene Skipping		*		*
Random Scene Accessing		*		*
Compression - Open/Closed	*	*	*	
Security	*	*	*	*
Encyphering	*	*	*	
keyed access	*	*	*	
cost	*	*	*	*
connectability	*	*	*	*
telephone interface	*	*	*	*
computer interface	*	*	*	*
ATM / Network	*	*	*	*

Figure 2.3 Distribution of ITV functions.

- Interactive games
- TV set controls
- Random scenes selection
- Billing review
- Cost control etc.

The one-button remote should be designed so that when it is picked up from a surface like a coffee table, menus appear on the screen overlaying the normal TV video. When the button on the remote control is depressed, a bright red laser-generated beam is emitted from the unit. When it is aimed at a simulated button on the screen, that button changes color, representing that this particular function has been selected. Inside the remote control is a photo cell connected to optics looking at the screen where the beam is pointing. The set top box will modulate the luminosity of each of the buttons with different low-frequency coded signals, which are detected by the remote and presented to the set top box. The one-button remote is an automated pointing device, a "mouse" of sorts, and could appear as shown on next page.

Menus. The menu system appears superimposed over the existing programming when the remote is activated. The menu appearances are different at different phases of program selection. For example, when the remote is first operated, the first-generation set of menus appears. When a selection from these is made, a second-generation set of menus appears, and so forth. If the remote is activated in the middle of a movie, for example, multiple scenes of the movie will be seen (up to 16) in a frame. Random view selection from any scene to any other scene can be accommodated by clicking on that small scene. This is analogous to fast-forward or reverse on a VCR, only faster selection and selection with full visibility is permitted.

The menus are designed to permit the display of:

- Program selection
- Scene selection
- Purchase choices
- Database selection
- Commercials
- Billing information
- Programming information
- Food services
- Other shopping services

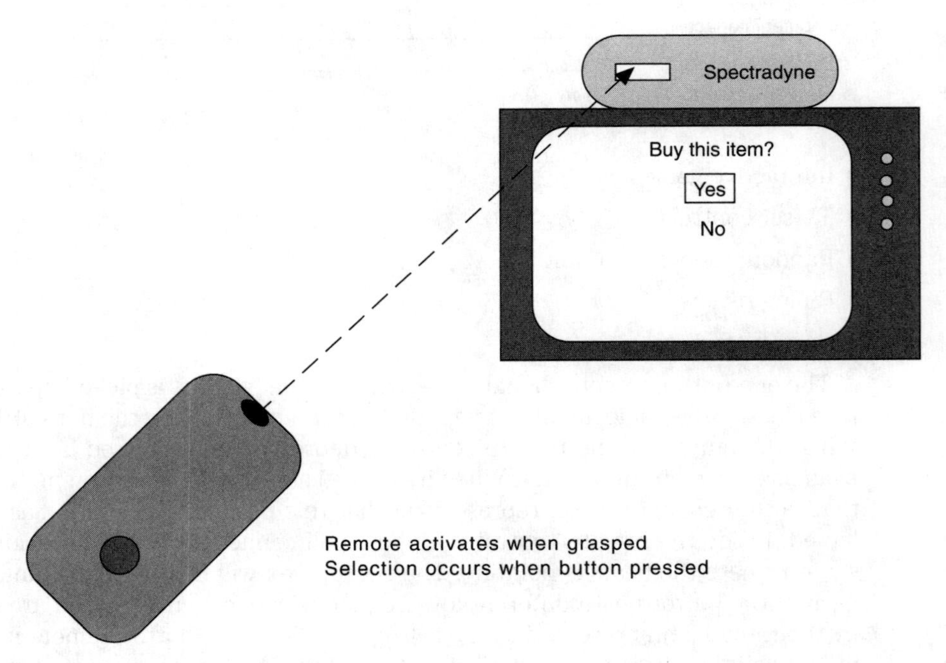

Figure 2.4 Intelligent remote control.

The menus are created at the CATV head end and are down-loaded to the set top box.

Navigation. When the remote control is activated, menus appear on the screen providing choices to be executed. A hierarchy of menus can (and often will) exist. The initial menu will provide initial selection to:

- Conventional TV
- Movies
- Sports
- News
- Shopping

Pointing to one of those five (or more) buttons produces the next set of unique menus. The conventional TV menu provides the number of buttons (icons) for the number of TV stations which can be selected. The movies button, when pressed, causes a menu to appear permitting selection of movie types (action-thriller, romance, comedy, tragedy, instructional, top 10, oldies, etc.). For example, selecting top 10 will create a new screen menu displaying buttons for the current top 10 movies. Selecting one of these movies causes the program sequence to begin. This process of selection via a hierarchical sequence of menus is called navigation.

Movie selection. Through the preceding navigation of menus, the selection of a specific movie can be achieved. The number of clicks of the one-button remote required to select a specific program may vary from four to 10, permitting the selection of a specific program from thousands of possibilities.

Interactive shopping. Shopping includes shopping for a movie, sporting event, or shopping interactively at an electronic mall. The menuing and navigation system permits one familiar and simple approach to select viewing material and/or product to be delivered with just a few points and clicks.

Interactive games. Games, whether team games or one-on-one games, can be played interactively with others. For example, playing chess with someone around the world, or with a machine, can be accommodated. The one-button remote would be used to drag the appropriate chess piece.

Object selection from the menu. The menu on the screen will consist of icons and photographs and/or small moving pictures that can be selected by pointing and clicking the remote control at them. When they have been clicked, the item they represent has been selected. If one of a series of small moving pictures is clicked, the film or scene representing that moving icon will be selected. If a picture of a pizza is selected, it will be delivered to the subscriber's door.

Windows. The icons, photos, and small moving-picture sequences are the objects of the window. These elements are superimposed on the existing TV picture.

Program interruption. The existing program can be interrupted by merely picking up the remote control. When a program is interrupted, a number of moving icons represent different delayed segments of the program, and an exit icon appears on the screen. Selecting an icon will cause the system to tune into a new delayed time slot for that program, or permit exiting the program directly.

Scene skipping. Scene skipping can be achieved by selecting moving icons for different scenes and clicking on them. When this occurs, the newly selected scene begins to play. The response time for scene selection or jumping in an NVOD system is practically zero!

Random scene accessing. Random scene accessing is identical to scene skipping, except the user selects any icon at random. Because a VHS-quality program may have 15 start times from one disk on a video server at no extra system cost, 15 time slots (or delay times) can be continuously broadcast in an NVOD system permitting a 90-minute movie to have 6-minute (i.e., 90/15) start times (or threads). Picking up the remote control permits the set top box to display all 15 threads simultaneously on the screen, permitting the viewer to select any thread (or scene) he or she desires.

Compression. It is convenient to represent an NTSC TV picture in pixels. There exists 525 lines (262.5 odd-numbered lines and 262.5 even-numbered lines, interlaced) every $\frac{1}{30}$ of a second. In North America, only 485 of the scan lines have picture content. Resolution within a scan line varies, but it is usually adequate to distinguish 300 vertical lines. Distinguishing 300 lines requires 600 pixels. Thus a 600 by 485 pixel display (VGA supports a 640 by 480 format) with 24-bit color requires:

$$600 \times 485 \times 24 \times 30 = 209{,}520{,}000 \text{ bits per second.}$$

Assuming QAM facilitates the encoding of 4 bits per cycle (hertz, or Hz), then 52,380,000 Hz of signal bandwidth would be required, which is clearly 10 times the bandwidth required by analog systems. Therefore, image compression is necessary just to maintain even ground. To maintain a conventional TV bandwidth, a digital compression factor of 10:1 would be required.

Most images consist of significant redundant information, and techniques have been developed to remove these redundancies. These are referred to as image compression algorithms. Frequently, these algorithms do not produce the desired results. These undesired results appear as artifacts or alterations to the images, which reduce the compressed image correlation to the original. However, image compression technology has advanced to the point where one high-quality NTSC image can be produced by a 7MB/s data stream. Assuming 4 bit-per-hertz QAM encoding, a bandwidth of 1.75 MHz is required, permitting 2.8 program sources to reside in the bandwidth currently required by one conventional NTSC program.

VHS-quality video can be produced by a 1.5MB/s stream, permitting thirteen program channels in the space of one conventional NTSC channel. Therefore, digital compression can produce a range of analog compression factors from 2.8 to 1 for high-quality video, all the way to 13 to 1 for VHS-quality video (as compared with conventional analog schemes).

Various compression schemes include JPEG, MPEG, wavelets, fractals, and spatial correlation technologies. These will be discussed in chapter 5. Hardware mechanizations that permit the changing of their compression algorithms are called OPEN compression architecture schemes.

Because computers can be used for viewing TV, and a TV can view computer data, an interesting comparison of screen resolutions is shown in Figure 2.5.

Image decompression. Image decompression is undoing the work done by the compression algorithm. It must be done in real time to provide smooth, high-quality moving pictures. It must be done inexpensively, because tens of thousands of decompressors will exist for every single compression CATV head end in the digital TV application. An open architecture is preferred for the compression/decompression algorithms, because the digital video compression science is in its infancy. The ability to download new decompression algorithms to the TV set top box for execution is desirable to preclude early set top box obsolescence.

Real addressing. Two types of addressing should be used. Real addressing mode should provide the set top box physical address. When the system needs to communicate with the set top box specifically, real addressing is employed. When the set top box communicates with the system, again real addressing is used. When a set top box requests a digital program, it requests the program using its real address. The system addresses the set top box using its real address, and instructs the set top box to use the accompanying virtual address to decode the requested program. A specific program is defined by its concatenated virtual address and its time slot address. Of course, a requirement for decoding a program is not only knowing the virtual address, but also the deciphering codes.

Virtual addressing. The temporary address used by the set top box to decode the specific program.

Security. The requirement to protect program material from usage by nonpayers or nonauthorized persons.

Format	Birth	# Pixels Per Line	Lines	Total Pixels	Frames /Second	Pixels/sec	Bits	Bits/sec
MDA	1981	720	350	252,000	50	12,600,000	1	12,600,000
CGA	1981	640	200	128,000	60	7,680,000	4	30,720,000
EGA	1984	640	350	224,000	60	13,440,000	6	80,640,000
VGA	1987	640	480	307,200	60	18,432,000	6	110,592,000
NTSC	1951	600	485	291,000	30	8,730,000	24	209,520,000
SECAM		580	575	333,500	50	16,675,000	24	400,200,000
PAL		580	575	333,500	50	16,675,000	24	400,200,000
8514/A	1987	1024	768	786,432	43	33,816,576	8	270,532,608
S-VGA	1988	800	600	480,000	72	34,560,000	8	276,480,000
XGA	1989	1024	768	786,432	43	33,816,576	8	270,532,608

Figure 2.5 Comparison of screen resolutions.

Enciphering. The process of encoding the program data so that it cannot be decoded without an appropriate authorization key.

Keyed access. The key to unlock the specific program.

Cost. A set top box should have a retail cost of less than $150.00 with no sacrifice of function.

Connectability. A total system that permits connection of set top boxes and computers via specific connectors on the set top box (or separate ATM cards that plug into computers that connect them to the network) should be provided.

Telephone interface. An optional RJ11 (standard tip and ring telephone jack) on the back of the set top box, which permits access to the worldwide telephone network via the CATV system.

Computer interface. A connector on the set top box that permits direct connection to the computer, like an Ethernet port, RS422 port, SCSI port, or serial Fire Wire port.

ATM network. The CATV digital network portion of cable services occupying the frequency spectrum from 500 MHz to 700 MHz or 1 GHz, with a reverse channel frequently in the 5- to 45-MHz spectrum interval.

ADSL (asymmetric data subscriber loop) network. The telephone companies' alternative to cable, using existing twisted pair wire to bring 1.5MB/s to 6MB/s digital program material to the subscriber. The set top box for ADSL systems will require an auxiliary daughter logic board.

Advanced multimedia. We have seen what video on demand can be. In this section, we will discuss advanced multimedia, compare it to its companion (video on demand), and compare it to 1990 era multimedia.

1990s-era multimedia is characterized by blurred images, fluctuating colors, slow frame rates, inability to provide satisfactory operation when shared over a network, and slow interactivity. The limitations are due to video-unfriendly disks, inadequate real-time compression, low-performance networks, and the absence of video servers.

Advanced multimedia incorporates video-friendly disks, adequate real-time compression, high-performance ATM networking, and modular video servers. Advanced multimedia can share ITV resources, functions, and performance characteristics, in addition to providing other functionality.

The subsequent block diagram shown in Figure 2.6 represents a conceptual composite ITV/advanced multimedia system. The connectability assumes that the data traffic between workstations is interoffice rather than intraoffice. An intraoffice system could be private, or it could attach to the public network via routers, which would be employed to keep local traffic local.

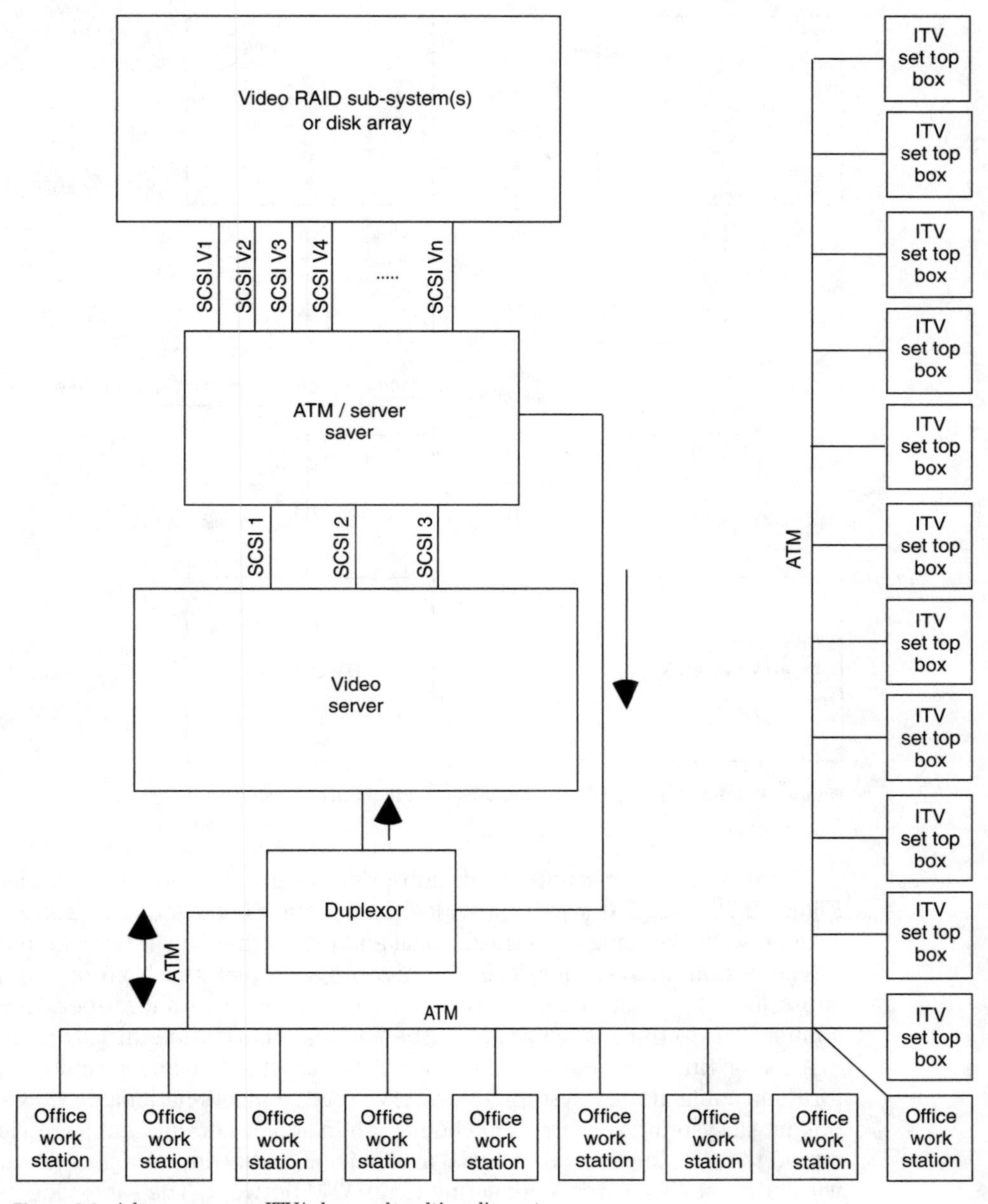

Figure 2.6 A homogeneous ITV/advanced multimedia system.

An advanced multimedia system is just as likely to be a private system with a private video server, private ATM network, and a group of personal computer workstations. It is identical to the drawing above, except all stations are probably workstations and no TV set top boxes would be employed.

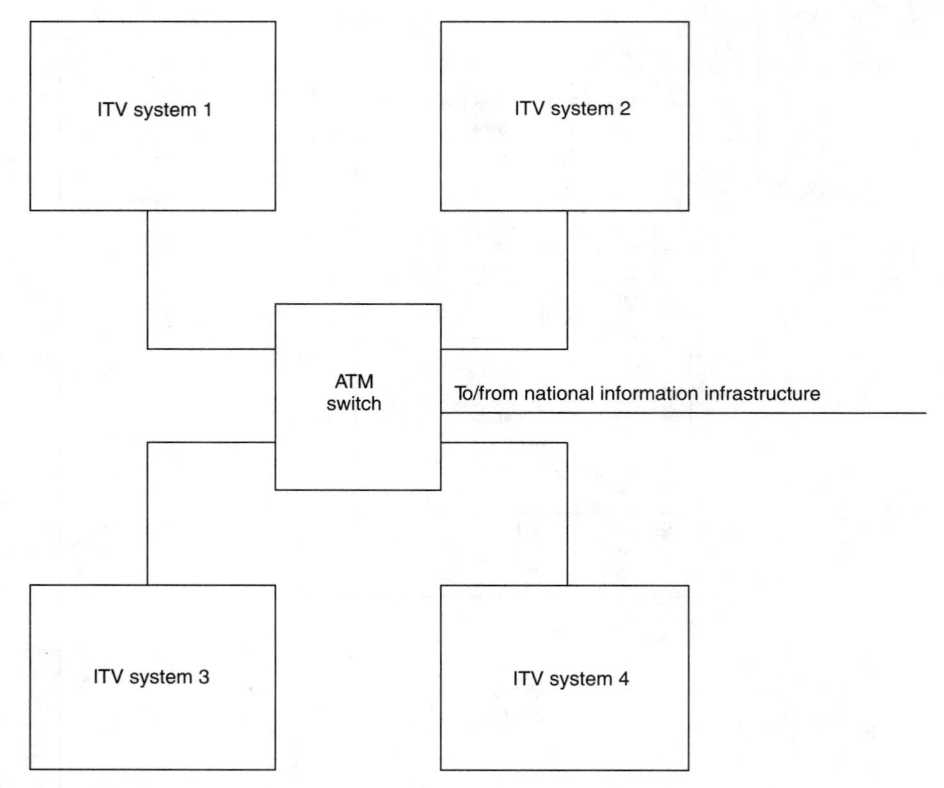

Figure 2.7 Interconnected ITV and/or advanced multimedia systems.

These ATM networks and their subscribers can be interconnected, as shown in Figure 2.7. The ATM switch provides the function analogous to a router in a local area network, keeping local traffic local and permitting only intersystem traffic.

The system above, therefore, permits program material from one system or its subscribers to be interchanged with another system or its subscribers. It also shows connections to the entire nation via the national information infrastructure.

A major European country currently is planning an advanced multimedia system for their public library system, to preserve that country's national heritage. They are building a system to archive 3000 hours of film and culturally significant videos. They intend to make any of that material available on audio/video Macintosh workstations with internal ATM cards, connecting to an ATM network. The same video server and ATM network technology will be employed as described for the ITV system in this chapter. Three modes of library patron operation will be permitted. They are:

- Library patron selects desired programming and views it from beginning to end.
- Library patron selects desired programming and views some or all of the material, with the ability to move forward or backward in the program and re-review program material.

- Library patron selects one or more programs, randomly accesses portions, records video pieces on his local hard video-friendly disk, creates lists of where the program pieces reside, creates an edit list, creates titles, and records new program sequences on analog or digital tape. This latter function incorporates full video editing capability. The final output is new programming material, which can be removed from the library and transported to any destination.

The only difference between this European advanced multimedia system and the ITV system is the level of interactivity and editing capability.

It is not my intention to go into great detail about this European advanced multimedia system; rather it is presented in this chapter to provide a summary overview of a truly advanced multimedia system, and how it is so similar to ITV systems presented elsewhere in this book.

The European advanced multimedia system is shown in Figure 2.8 below.

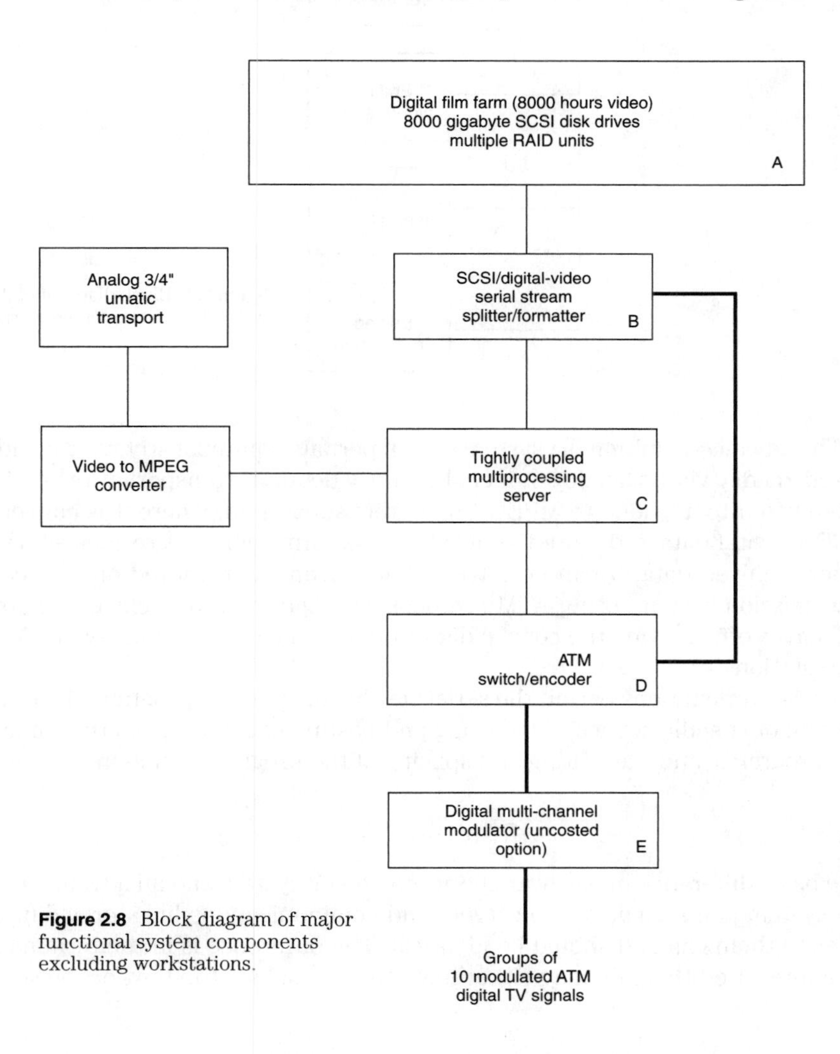

Figure 2.8 Block diagram of major functional system components excluding workstations.

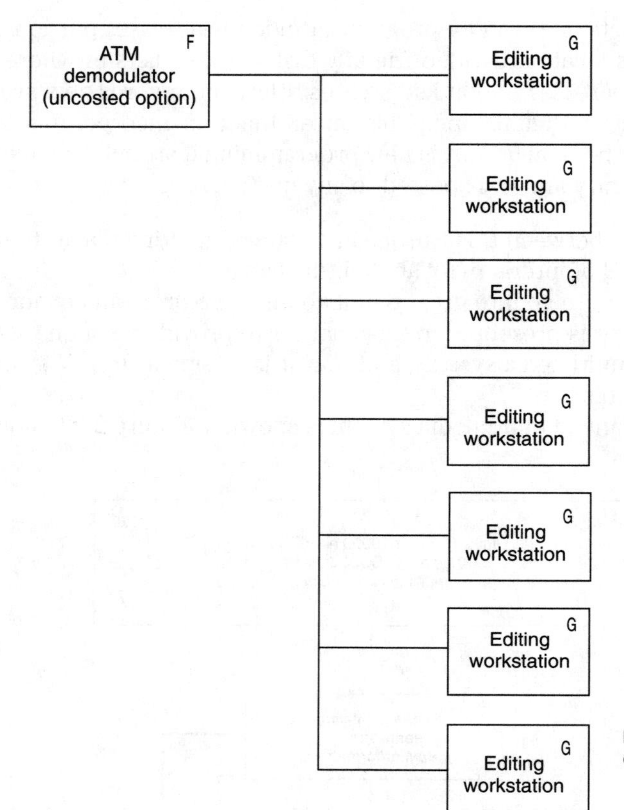

Figure 2.9 Block diagram of ATM channel and audio/video workstations.

The advanced multimedia system shown permits continual archiving of videotape-based media via an analog ¾-inch Umatic videotape transport. Conventional TV video is fed to a "Video to MPEG Converter" subsystem, where it is converted into MPEG digital data and stored on a "SCSI disk farm" where it resides as part of the other archived data. It can be retrieved at will and transmitted on a coaxial cable transmission system, using ATM protocol. The figure above picks up where Figure 2.8 leaves off, showing the coaxial distribution to a cluster of audio/video Macintosh workstations.

These workstations permit the serial viewing of an entire program, the random accessing of specific scenes, the cutting and pasting of scenes from the same or various programs, and the editing and splicing of the program segments.

Conclusions

The basic difference in hardware between an ITV system and an advanced multimedia system is marginal. Video servers and communications links providing the services to them can and should be identical. The difference is in the terminals. It can be appreciated that ITV may either have an external set top box, or the set top box functions can be internalized within the TV set; whereas advance multimedia replaces the ITV set with an audio/video-equipped personal computer. The ITV relies primarily on the one-button remote control for navigation, while the advanced multimedia workstations or personal computers employ a keyboard and mouse to provide the same function. Editing can be more easily performed on the personal computer, and the personal workstation user could watch ITV programming and interact accordingly on his PC. What is the difference? The technology is *converging.*

Chapter

3

Video on Demand: Architecture, Systems, and Applications

Note: This chapter was previously published in the September 1993 Issue of the SMPTE Journal by Winston Hodge, Stu Mabon, and John T. Powers, Jr.

The authors of this chapter have been actively involved in computer architecture, imaging, storage systems, transmission systems, video on demand (VOD) and pay-per-view (PPV) TV systems for many years. They have been direct participants and have followed the emerging technologies that are changing the fundamentals of telecommunications, transforming telephone and CATV companies into comprehensive communications deliverers that will integrate the functions of today's telephone, cable television, videoconferencing, and data communications networks, and which will provide for direct competition or partnering between TELCOs and CATV companies. With the advent of cellular TV, the possibility of video stores becoming local VOD television delivery services looms on the horizon. Another opportunity for small entrepreneurship will be born. The same technology will also apply to hotel VOD/PPV television systems. This chapter will address a top-down implemented hardware/software architectural solution to PPV/VOD that is modular and scalable, so as to bridge systems from 50 to 50,000 subscribers and from 20 to 2000 program sources in a cost-effective way. A common solution permitting growth will be shown to exist for small and large companies alike, and will be described in this chapter. Our chapter will discuss:

- VOD system requirements
  - —TELCO
  - —CATV

—Cellular TV
—Hotel PPV systems

- Open systems architecture requirements
- System availability
- Cost philosophy
- Operation support
- Scaling and modularity
- Network dependencies
- Hardware
- Software
- Subscriber equipment
- Program database content
- Subscriber program search/sort facilities
- Subscriber interface
- Relevant patents

Asymmetric Model of Information Consumption

Information systems that communicate directly with humans are characterized by a large difference in the amount of data transmitted versus the amount of data received. The human eye/brain combination is capable of absorbing a very large amount of information quickly, while the hands that operate keyboards and positioning devices are many orders of magnitude slower. To ignore this difference is to risk producing uneconomical and even unusable systems. Consider simple television viewing using a remote control; while current technology requires megabits per second to deliver home-quality TV, the average data rate from the remote control to the TV is in the order of a few bits per minute.

This asymmetry is a relatively new concept in the computer and data communications fields. High speed computer channels, local area networks, and both analog and digital leased lines have all been designed with the same data rate in both directions for their main transmission paths; this is based on the assumption that data exchange will be between computers. The entertainment business, on the other hand, has used asymmetric communications for years; one example is pay-per-view TV, where a program is ordered simply by dialing 11 digits on the telephone. The ratio of user input (44 bits to code an 11-digit telephone number) to response (say 90 minutes of video at 192,000 bytes per second) is over 188 million to 1. Video applications make significantly different demands on digital storage and transmission systems than those for which they were designed. This chapter will examine some of the principal issues in interactive video system design.

The program path

By "program path" we mean the path by which video information is delivered to the viewer. While many technologies exist that are capable of transmitting video, all of them share the characteristic called "high bandwidth"; compared to other applications, they require a great deal of transmission capacity. This is true whether the transmission is analog or digital and is simply a result of the amount of information delivered per unit time. For example, an analog TV channel requires about 1500 times the bandwidth of a telephone call.

The return path

The transmission path by which the user communicates his or her requests for programming to the video storage and transmission system is called the return path. This is consistent with cable television terminology, in which (for 2-way systems) the television signals are sent in a "forward" direction and signals from decoder boxes, telemetry systems, etc. follow a "return path." The originator of information sent along this path is a person, most likely using a conventional-appearing TV set top box, a keyboard, or a pointer of some sort. Some uses for the return path are:

- Program selection
- Response to polling (as with an electronic town meeting)
- Answers to quiz questions presented by an educational program
- Browsing and ordering from an electronic shopping catalog
- Real time communications analysis via return telemetry data analysis

The return path can be: (1) the same cable, (2) different cables of the same type, (3) different cables of different types, (4) a different return frequency or (5) a combination of the above. The return path can have a different owner than the program path.

Trends in video technology

While all aspects of the video business are involved in revolutionary change, certain trends are easily identified:

- Digital capture and storage
- Digital transmission
- High definition
- Wide aspect ratio

Digital video storage on magnetic tape has been in use by broadcast and post-production houses for several years, using products from Ampex (e.g., "D2" format) and others. As compression technology and disk price/performance improve, random access storage will become the rule rather than the exception for storage of active (nonarchival) material. Digital capture (broadcast-quality digital cameras) are not far behind, waiting mainly for cheaper high-density versions of the CCD arrays already in use for consumer camcorders. All-digital TV studios, control rooms, and editing facilities will be old news by 1995, capturing very-high-quality images and sound, reproducible essentially forever without degradation.

Digital transmission is now taken for granted in the cable TV industry, with the only questions left being implementation details and schedule. In addition, certain regional Bell operating companies (RBOCs) have also become missionaries within the U.S. telephone industry for digital transmission over existing local cable plant, using a system called asymmetric digital subscriber line (ADSL).

The public, used to digital music recordings and telephone calls, seems quite ready to associate "digital" with "high quality" and "high function" for video after over 40 years of living with TV in its original form. It is almost certain that the next U.S. TV standard will be not only much higher in resolution, but completely digital.

Aspect ratio is the ratio of picture width to height, and the U.S. advanced television will most likely have a 16:9 ratio. Digital storage and processing immensely simplifies the problems of automatic format conversion, not only for aspect ratio, but for the differing frame rates and resolutions of TV systems around the world.

Economic Trends

The cost of electronic hardware, especially that of mass-produced items like memory chips, continues to plummet at an amazing rate. Magnetic storage prices are also falling rapidly, but not as rapidly as for semiconductor products.

On the flip side, the price of human talent for video production and post-production continues to rise. Technology that amplifies human talent and increases the accessibility of products like high-quality television productions will gain increasing acceptance.

Regulatory Trends

Government regulation seems to be waning all over the world, with the result that limitations on business activities are being relaxed. In the U.S., telephone companies, particularly the RBOCs, are gaining freedom to sell information as well as carry it. While it remains to be seen whether or not they can rise to the challenge, the potential exists for major competition to newspaper, radio, and television businesses from cash-rich telephone carriers. The trend toward "pay" vs. advertising-supported mass media will gain speed, and the media industry as we know it may vanish in the next few years.

System Requirements

The various pay-per-view VOD systems share many characteristics, but differ primarily in their transmission system and return path technology. The telephone companies, for example, using ADSL technology, can transmit over 1.5 MB/s over conventional copper pairs to their subscribers. Using video compression technology such as MPEG, they can deliver VHS or better quality television to the subscriber. Because of the telephone company's' copper wire transmission topology, it could be possible to send a different program to each subscriber. On the other hand, CATV companies typically have a transmission capacity of 35 to 70 conventional analog television channels on a delivery cable. But the CATV companies are also exploring

digital transmission technology. By employing image compression technology, CATV companies can expect to achieve a 16- to 24-to-one increase in utilization of their facilities, permitting the transmission of in excess of 1300 VHS-quality MPEG-encoded video television programs concurrently. The chart below shows the relative capacity of 6-MHz channels when VHS-, broadcast-, and studio-quality video are encoded assuming 6 bits per Hz encoding.

MPEG quality	MB/second	# programs / 6-MHz channel	# programs / 500 MHz
VHS	1.5	16	1333
Broadcast	5.0	5	416
Studio	7.0	3.5	291

The next table illustrates encoded TV capacity when a more conservative 4 bits per Hz encoding is employed. The two tables represent an interval of realistic expectations.

MPEG quality	MB/second	# programs / 6-MHz channel	# programs / 500 MHz
VHS	1.5	10	880
Broadcast	5.0	3	274
Studio	7.0	2	192

An interesting parameter of this architecture is that it could support concurrent variations of VHS, broadcast, and studio quality at different costs to all subscribers. Because there exists 83 six-MHz channels in 500 MHz of spectrum, and assuming 6 bits per Hz encoding, as many as 291 studio-quality programs, 416 broadcast-quality programs, 1333 VHS-quality programs, or some combination thereof could be provided by the system. Subscribers would pay more for higher-resolution programs.

The FCC has recently approved the frequency range 27.5 to 29.5 GHz for experimental cellular TV. It is general knowledge that experimental frequency allocations that are successfully used become permanent. The purpose of cellular TV will be to provide local PPV TV with transmission ranges of no more than a few miles. It is viewed as a CATV alternative for subscribers. Because the range of the cellular TV is probably comparable to the geographic range for video rental stores, and because the investment in facilities may be less than $250,000, an opportunity for small businesses to compete with CATV will exist. Cellular TV will have comparable capacity to CATV systems.

Hotel PPV VOD systems are functionally identical to TELCO, CATV, and cellular TV except the choice of transmission facility can be the hotel's private TV cable system, twisted copper telephone set to PBX pairs, low-power transmission in the 27.5- to 29.5-GHz range, or some combination.

Program transmission paths for VOD could be conventional TV NTSC broadcast approaches, but insufficient capacity and flexibility for that system can be realized. Four system technologies exist that will have the capacity to provide good quality and adequate program sources. These systems use digitally compressed video and audio signals sent over subscribers' (1) telephone lines, (2) CATV systems, (3) hotel pay-per-view systems, or (4) small-scale cellular TV systems. Digital compression will permit multiplying system capacities by an order of magnitude over conventional broadcast TV. A VOD system requires a return path so that a specific program can be selected by the subscriber from a choice of many programs. VOD generally implies that a video program can be seen within 5 minutes after it has been selected. Some VOD proponents mandate instantaneous video program response. While that is doable, it is expensive, and the extra capability cannot normally justify the cost. Similarly, the VOD purists believe the subscriber should have the same control over his selected video transmission as he would on a tape in his VCR, such as fast forward, reverse, pause, etc. These features are also achievable with the architecture we present, but at a considerable cost. We believe that when it becomes necessary to stop the video (pause), restarting on 5-minute intervals will not cause any inconvenience if a suitable intermission program is available for that duration. Similarly, immediately after requesting a video, an opportunity will exist to present video segments of pizza from Domino's or other fast food delivery services, or short advertising features meeting the profiled interests of the subscriber so these products or services can be electronically ordered. Transmitting video in 5-minute increments significantly reduces the number of program threads, and therefore also reducing the video server seek time and channel loading requirements. Going to the video store takes probably 20 minutes to drive to the store, rent the video, return, and set up the VCR. After the video has been watched, the videotape must be returned to the video store (total lapse time = 40 minutes). It can be seen that a return path to request a selected video is required. Some people would refer to our VOD technology as *near video on demand* (NVOD).

The VOD return path can be a telephone circuit, a reverse path on a CATV system, or a polled radio path for the video program selection processes. Figure 3.1 represents a generalized VOD system. At the subscriber facility is a means to request and view the selected video. It could be an integrated set top box providing all the necessary functions, or it could be less glamorous and require the customer to phone (or even mail) the request, or it could record internally the viewed program billing information for subsequent collection and billing purposes (someone has to read the meter). Our VOD block diagram representation (Figure 3.1) is independent of the transmission facility, and the transmission facility only limits the number of distinct concurrent programs and the quality of the video. The return path is one of the facilities that define the system.

System Definitions

A TELCO system. It transmits a video program over existing copper twisted pairs (used also for plain old telephone service—POTS) using ADSL and MPEG technology. Encryption, a procedure designed to limit reception to intended parties only, is not necessary on the telephone lines because a video is not transmitted to the subscriber until his or her request is validated, and then it is placed on a customer-specific copper pair.

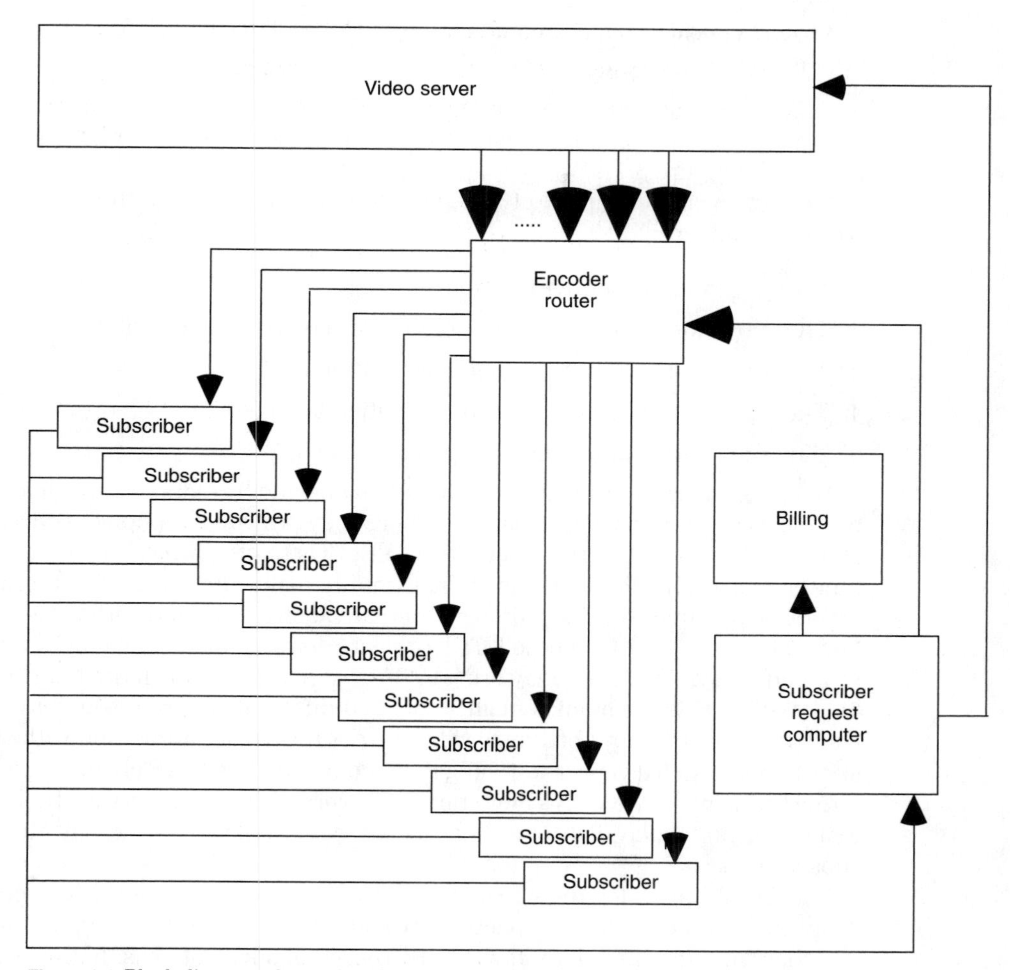

Figure 3.1 Block diagram of generic video on demand system.

The CATV system. It uses the conventional TV cable to broadcast the encrypted video to the subscriber. Encryption is required because all signals are sent to all subscribers. A subscriber's key permits unique decoding of the program. The return program selection path may be (1) the cable, (2) a telephone line, (3) a postcard, or (4) retroactive meter reading from the set top box.

The cellular TV. Implementation is the same as the CATV except it may have a radio signal for a return program selection path.

A hotel system. This is a system designed to deliver VOD to hotel guests. It may use the hotel coaxial TV distribution system, existing in-house POTS copper phone wiring, or low power RF signals.

A hybrid system. This is a combination of CATV, TELCO, cellular TV and/or hotel system facilities, and may use fiber optics.

All the systems require a video server, an encoder/router, a subscriber request computer, a billing computer, and the set top boxes at the subscriber location. This is shown in Figure 3.1.

The significant elements of the generalized system represented by the above block diagram are:

1. Set top box at subscriber location
2. The subscriber request computer at the video origination facility
3. The billing computer at the video origination facility
4. The encoder/router at the video origination facility
5. The video server at the video origination facility

Telephone companies (TELCOs) have been barred by federal law from owning cable television systems in the markets where they provide telephone service. Early in 1992, the FCC adopted rules permitting TELCOs to offer what is called "video dial-tone" services for other companies that want to distribute television programming.

Cable companies routinely offer more than 50 channels, and with the event of the high-capacity CATV GI/ATandT/TCI project, delivery of 550-channel systems are well underway. The future may bring CATV systems delivering more than 1600 channels. While cable companies might have 550 different program sources, it is more likely that some (many) channels will transmit the same information with staggered start times, to make viewer watching more convenient (e.g., 18 channels could be offering the same movie with a new start time every 5 minutes). On the other hand, the TELCOs using video dial-tone theoretically could deliver a different film to every subscriber!

Cellular TV provides the capability to reuse frequencies from cell to cell just as cellular phones do. The RF spectrum would be divided up between multiple video providers in different cells and would be reused in adjacent cells. In this way, different cellular TV companies can compete for the same and different local areas.

The provision for anywhere from 20 to hundreds of program sources has been facilitated by digital video compression, video servers, and video-compatible large SCSI disk drives.

System Requirements for a Video on Demand System

The authors have suggested the following requirements for PPV VOD systems.

- Subscriber populations of 200 to 75,000 need to be supported.
- Distinct customer-selectable program sources from 20 to 2000 should be supported.
- This system must serve 100% of all subscriber user sets simultaneously, at any point in time.

- This system must permit real-time interaction between the subscriber and the system.
- Basic services to be offered to subscribers include video entertainment programming, video educational programming, interactive games, mixed-media information services, and shopping services.
- The system must have basic navigation functionality that permits a subscriber to make a choice of services within the video retailer's system. The navigation system must be simple to use, and a graphical user interface (GUI) similar to Microsoft Windows, IBM's OS/2, or Macintosh System 7.5 should be considered.
- More than one interactive subscriber device must be supported in each dwelling unit.
- Open compression architecture should be pursued, because this will give the video retailers an orderly growth plan into the middle of the next century, providing higher-quality video as better compression algorithms evolve, and providing compatibility with HDTV formats as well as NTSC and others. The algorithms should permit decoding the same digital video programming differently, to accommodate TV set variations in resolution and aspect ratio.

Generalized VOD System Architecture

VOD systems require (1) an information source, (2) video storage and distribution facilities, (3) program switching or routing, (4) a management facility that communicates with the customers' set top boxes (to determine what program sources are available, what they cost, and coordinate which customer gets what program), (5) the set top box at the subscriber end, and (6) the communications facilities that permit program selection.

The block diagram of the VOD system in Figure 3.2 considers how video resalers shall obtain their video sources. Sources may include motion pictures, interactive video (education, marketing, games, sales), remote broadcasts (BBC, Singapore TV, CBC, etc.), . . . and the list goes on. The subsequent top-level overview diagram represents the universe of VOD, showing numerous VOD delivery companies. While one video wholesaler is shown, numerous video wholesalers are expected to exist, wholesaling different types of material. Industry rumors indicate that companies such as Next Century Media, EDS, Kodak, HCR, et al. may be contemplating relationships to provide this top-level function. They would deliver video to the video retailers electronically. The video retailers will provide storage and distribution facilities to deliver the product via their wires (or over the air via cellular TV) to their customers. Billing provisions will be established to permit either the video retailer or the video wholesaler to provide end-user billing.

The subsequent drawing therefore represents the entire path—the video wholesaler, the subscriber delivery service (video retailer), and the subscriber.

This chapter will spend the rest of its attention to the local distribution and subscriber areas. This is shown by the following system block diagram.

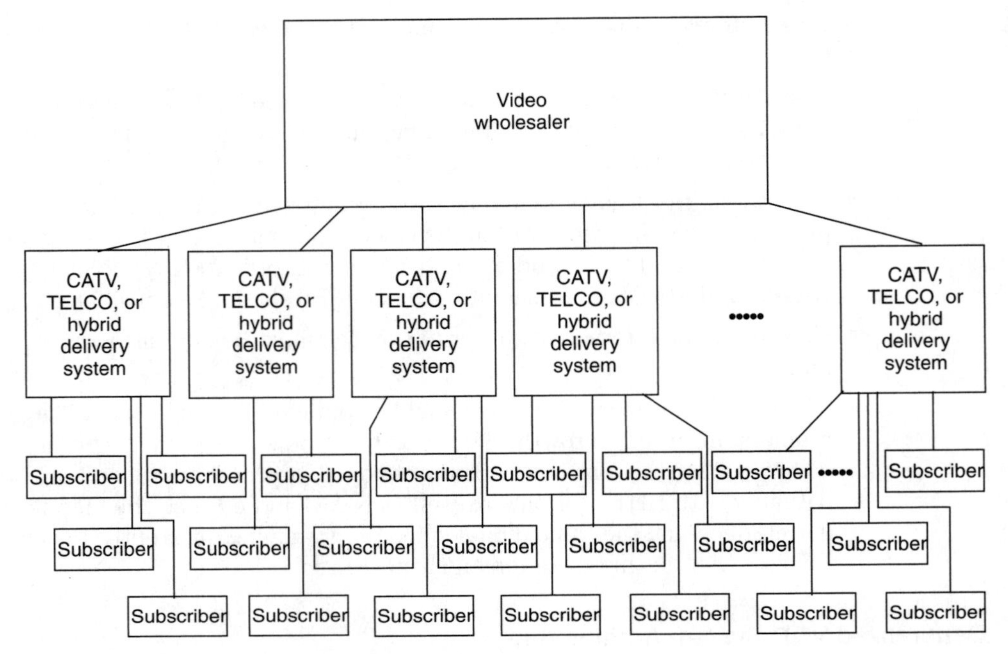

Figure 3.2 Macroscopic view of the VOD system universe.

Common building blocks

A generalized pay-per-view (PPV) VOD system could appear as shown in Figure 3.3. It is important to analyze a VOD system architecture that shares components with as many different delivery systems as possible to determine if real commonalities exist, and if economies of scale can help drive down system costs. There probably would not be much argument that at least the video server block could be compatible with all delivery systems. When we look further, the billing facilities could be identical. Further, the digital image and audio compression could be the same. The philosophy for program selection by the customer could be the same, and therefore the program selection computer could be compatible for all delivery systems. The bottom line on differences is how does the signal get to its destination, is it multiplexed or a single signal, is it a space division technique or a time division technique, and is it wired or broadcast as radio waves?

Because the video server is probably the most important building block, our discussion will start with that.

Video server

VOD PPV system operators require storage and retrieval systems for their VOD service applications. This facility is referred to as a video server. Video servers are different from other computer servers because of the vast storage required for motion pictures, and the high throughput requirements. Large mainframe computers can be used as video servers, but this implementation is costly. Servers designed specifically for video selection and distribution can be significantly more cost effective and provide the required performance and reliability.

The server must provide for random, simultaneous access to "top hits" of movies, music, interactive programs, and software. The server must also provide for sequential, batched access to on-line media such as digitally stored movies on disk or tape. It must be possible to distribute the inventory in the server on appropriate storage devices, memory, or physical media, so as to maximize the number of viewers and revenue flow.

The server must be designed for the graceful recovery or failure of application software or video storage. The server must also allow for the introduction of additional servers without network or software changes, and these servers should be "hot" pluggable while the system is in full operation. The server should provide redundancy to the extent that it is possible to recover and restore the server to full operation when component failures occur, without significant downtime on the network.

The VOD video server requires the hardware and software elements of a streamlined multiprocessor video database system. Such a system has evolved at Hodge Computer Research into a product designated "Super-Server." The system by itself is a powerful image database server, but, with the use of Micropolis Video SCSI disks presently under development, it becomes a massive high-performance but cost-effective distributed video database network. It is scalable and modular.

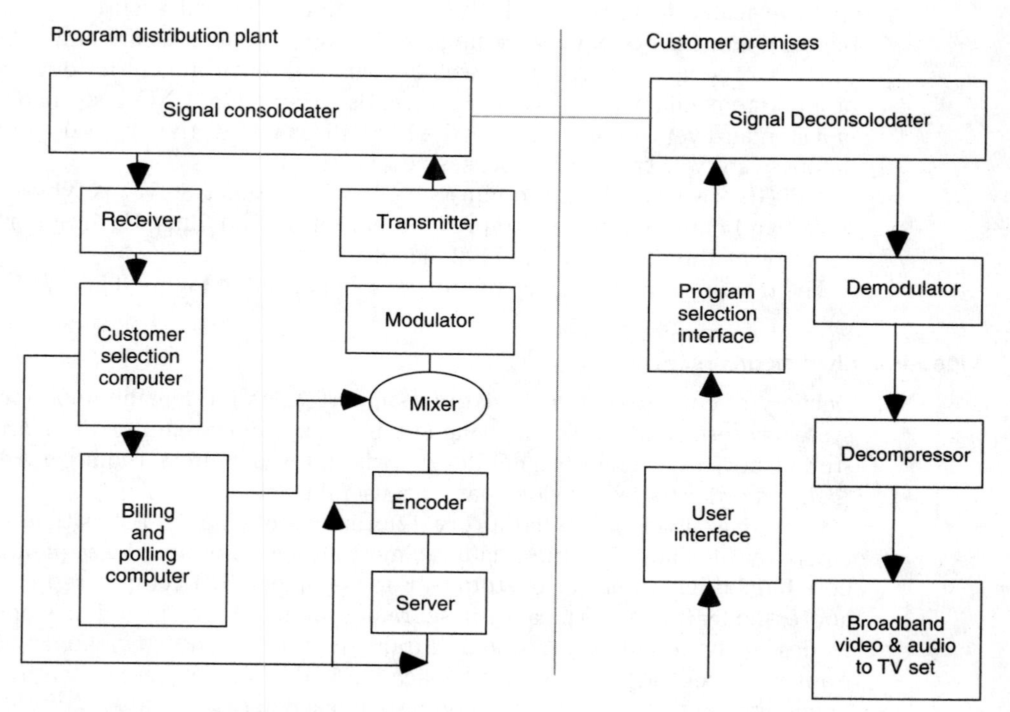

Figure 3.3 A generalized system.

The timing of this development was aligned with the release of the Intel i486 (and forthcoming Pentium) microprocessor. The use of this powerful microprocessor places the Super-Server in the front ranks of modern video database systems. Super-Server operates a standard version of UNIX with a multiple-processor option. This operating system directly supports a standard version of Oracle, as well as other applications that may be suitable for installation on the Super-Server nodes.

The support of multiprocessor UNIX requires a symmetric hardware architecture, with provision for minor specializations of certain processors to support scalability and modularity. This design provides for such a structure using Intel i486 microprocessors supported by multiple SCSI channels and appropriate network interface adapters. As many as eight i486 microprocessor-based CPU boards may be configured, along with 10 SCSI channels, delivering 60 digitally compressed video channels and 64MB of ECC-protected memory, to form a large, reliable, high-capacity single node.

The architecture of the Super-Server shown in Figure 3.4 provides a multiprocessor design that is designed to carry a standard Santa Cruz Operations (SCO) version of UNIX with the multiprocessor option. The primary storage within the Super-Server is provided by embedded Micropolis SCSI video-ready disk drives. These small, well-designed devices pack a large amount of storage in a very efficient package and provide video stream continuity not available from other high-capacity, high-performance SCSI disk drives.

This hardware structure is specifically designed to carry the standard version of UNIX available from SCO with the MPX option. This will ensure that the Super-Server operating system can be upgraded as new capabilities are added by SCO. It furthermore ensures a standard, well-known UNIX environment for development of other system enhancements in the future. Use of standard UNIX also enables the use of a standard version of Oracle for the RDBMS, as well as the direct upgrades to the database software as Oracle releases new versions.

A 20GB (twenty 90-minute film) version of the Super-Server is typical. Multiple units can be interconnected to provide multiples of 20 films. Each Super-Server is about twice the size of an IBM PS/2 model 80.

The relationship of these components is illustrated in Figure 3.5.

Video-friendly disk drives

Delivery of compressed video data to support video on demand and video editing systems presents a new set of challenges to the system integrator. Standard rotating storage devices, even intelligent devices with internal buffers, require a great deal of "help" in order to deliver video data in a useful fashion.

Compressed video data has a very different set of characteristics than traditional computer file data. Computer data is almost always addressed by sequentially numbered, fixed-length blocks of storage. For example, JPEG-compressed video data is addressed by frames. Typically, these frames are numbered by a time stamp that includes the hour, minute, second and frame number (standard video data has 29.97 frames per second). A further complication is that the length of each frame varies depending on the success of the compression scheme.

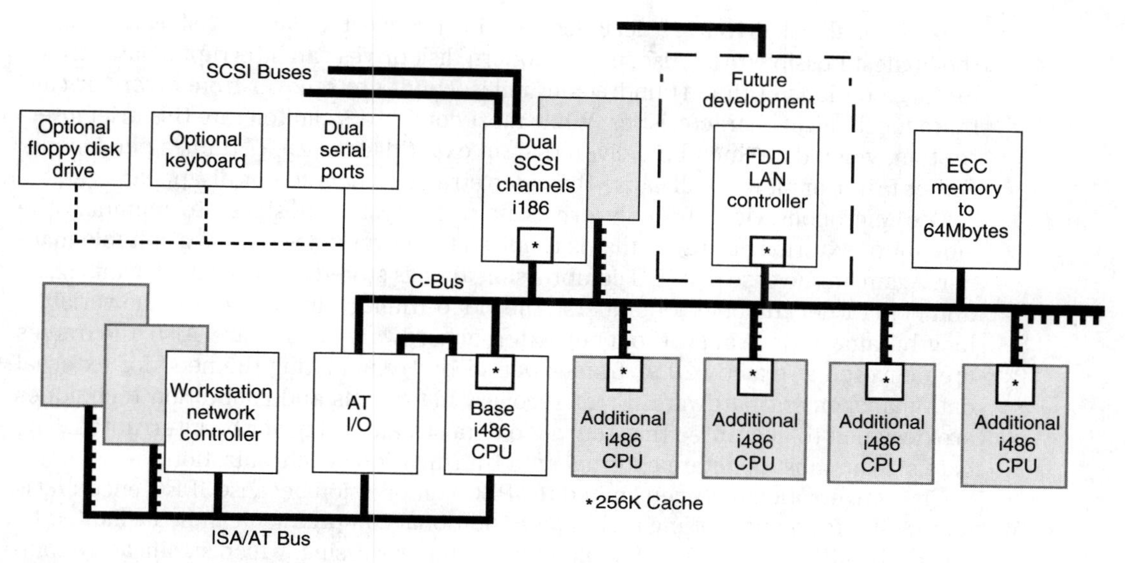

Figure 3.4 Super-Server node block diagram.

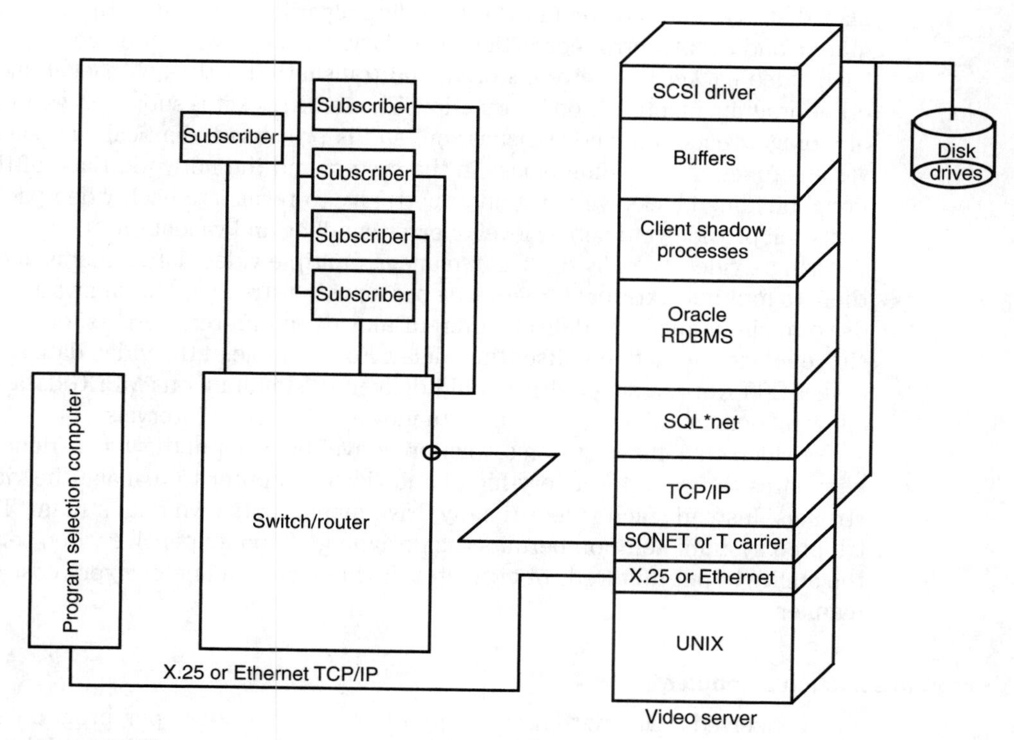

Figure 3.5 Video server and switch/router configuration.

Traditional disk drives are designed to deliver requested data as fast as possible at the highest possible rate. Also, many modern disk drives can interrupt a data stream for large periods of time (hundreds of milliseconds) to recover from errors or calibrate mechanical characteristics. While most computers can tolerate this brief interruption, video data must be delivered at an exact pace of 29.97 frames per second. Either faster or slower will cause the video stream to degrade or disappear.

The Micropolis video-friendly drives help the system designer to minimize the amount of external hardware that is required to deliver video data in a suitable manner. Again, in the case of JPEG compression, data is stored and retrieved using frame numbers rather than block numbers. The video-friendly drive manages the variable-length frame addressing without any external hardware or software. Also, the frames are delivered at exactly 29.97 frames per second, eliminating the need for external data management hardware. Error recovery algorithms and calibration techniques are designed to guarantee that the video data stream will never be interrupted. The data stream must facilitate external data transmission synchronization.

The conversation has been limited to JPEG compression because JPEG encoding is a frame-by-frame processing technique. Additional compression can be facilitated by one of the MPEG variants by including temporal processing, which significantly complicates frame addressing. Similarly, adaptive polyhedral encoding algorithms exploit intra-frame redundancy and suffer from the same unique frame selection complexity. A video-friendly drive will accept, at the beginning of the operation, a disk formatting plan that is optimized for (1) the encoding algorithm used, (2) the resolution required, and (3) any error correction necessary. The error correction code attached to each video packet is received, stored, and transmitted with each packet, along with other program identification information. The same packet is subject to various stages of storage, reception, and transmission, and is processed identically by each of the various functional building blocks in the system and the network. Each of the functional building blocks will have an opportunity to revalidate each video packet. The disks will provide yet another level of error handling and validation.

A single video-friendly drive can manage multiple video data streams and deliver them to multiple external targets as separate video streams. This allows a single disk drive to simultaneously deliver different movies to different targets, or to deliver a single movie with a fixed offset time delay. For example, if the video data stream averages 200 Kbps, a single drive could deliver 18 simultaneous MPEG data streams, allowing delivery of a single 90-minute movie at 5-minute intervals.

For video on demand systems, this "proactive" behavior of the video-friendly drives eliminates the need for a very-high-bandwidth computer to manage the video data streams. Instead, each video-friendly drive manages its own data stream. This "distributed system" solution permits the designing of an inexpensive video server with the ability to put hundreds of gigabytes of video data on line in a very cost-effective manner.

Program selection computer

Our conceived VOD system will support 6000 subscribers per program selection computer. Multiple program selection computers would permit multiples of 6000 subscribers to be added to the VOD system. Each program selection computer will have physical space to add up to 6000 logical 19.2 Kbps ports (200 ports per printed circuit card for 30 communications cards). The capacity of the program selection computer will permit simultaneous synchronous packetized communications. The program selection computer will provide a perception to the human observer via his or her program selection box of immediate or near-immediate response. Just as in multiuser time-sharing systems, when insufficient processing time exists to provide instantaneous response, simple questions, statements, and short video sequences will occur that permit processing to complete for other users. This will create the customer perception of an immediate response. From the time of selecting an item to receiving a new menu or program will not exceed 6 seconds. When the selection is complete, no additional delays will occur.

Traffic-induced delays can be caused by too many subscribers accessing the system at one time, and the bulk of the delay will be due to the program selection computer handling numerous transactions. For example, a video program that 4500 subscribers are watching completes at 8:00 p.m. At the end of that program, 4300 customers select a new program within that 1-minute interval. This represents 72 complex transactions per second if no preplanning is implemented by the subscribers. Our plan intends to employ a high-end workstation or comparable computer to service these requests as the program selection computer.

The VOD system facilitates control and programming of the storage and retrieval system by the program selection computer. The video server is loaded by the information wholesalers/information provider and unloaded by the *delivery service* (the TELCO, CATV, cellular TV, and/or hotel) under direction of the program selection computer. The subscriber selections are made to the program selection computer from their TV set top boxes (selectors).

TV set top program selection devices (selector box)

The *selector* (TV set top device) is that part of the VOD system by which customers communicate their current or future program choices. It is a part of the customer premises equipment (CPE) purchased by or provided to VOD customers. It may include a wireless remote control for customer convenience. It should be packaged so it can be sold by retail distribution outlets such as Sears, Radio Shack, telephone stores, specialty electronic stores, etc. The logical relationship of system functional components is as follows:

The selector function should be an intelligent interface between the customer and the rest of the system; in a sense, it is the customer's "agent" in dealing with both the network and the information provider. It should include microprocessor technology with a combination of fixed and downloadable program store, in order to offer maximum flexibility to facilitate an open systems architecture.

The VOD selector would, at minimum, embody the following functions:

1. Indicate the status of the customer premises equipment, network transmission, and program source, displaying the service company and/or information provider-originated messages and menus on the customer's TV screen.

2. Transmit customer choices to the serving office and/or information provider.
3. Provide customer control of basic functions, such as-power on/off and choice of VOD or standard TV operation. In addition, the VOD selector could provide the following functions, if not prevented by the design of other system components:
4. Control of customer's TV and/or VCR
5. Interactive shopping
6. Interactive education
7. Interface to home computers
8. Alarm transmission
9. Utility monitoring
10. Telemetry feedback of transmission performance

The last item above deserves special emphasis. Network environments are dynamic, compression algorithms may have differing error susceptibilities, correction schemes can be changed, etc. In particular, average error rates and the distribution of error burst lengths are unknown. Design of optimal error correction coding, however, requires this information. By gathering error statistics in real time and transmitting them to a central collection point, the performance of the system may be improved based on its analysis. If the error correction process at the subscriber end is under program control, such a performance increase could be achieved automatically by downloading revised control software. In this way, different error correction algorithms could be used in different geographic areas under different environmental conditions providing adaptive and enhanced operation.

The use of a microprocessor and stored program control, coupled with the ability to modify the program by downloading, is an extremely powerful way to guarantee the video delivery utility's ability to add or change system functions in the future. In a regulatory environment that is extremely volatile, these features will support rapid response to new market opportunities.

The *video decompressor* is logically housed with the selector function in a small box on top of the TV. The video decompressor function should decode 1.544 to 7 Mbps MPEG or other encoded video and convert it to raw or modulated (channel 2 or 3) video and stereo. The logic for the selector and the decompressor functions are likely autonomous, except for on-screen menus of program and teletext information. Because 1.544 Mbps MPEG renders VCR-quality video, the video decompression architecture should be open, permitting use of higher transmission rates and execution of better video decompression algorithms as they become available, to permit the transmission of HDTV-quality video without obsoleting the program selector box or the VOD facilities at the video delivery utility.

Programming Compression Schemes

The authors are proponents of *open architecture*, which will permit hardware to withstand long periods of time. Just as NTSC has lasted 50 years, so should digital TV.

MPEG and its variations suffer from undesirable artifacts, variations in available resolution as a function of the bit stream rate and amount of motion, the inability to provide concurrent variations in aspect ratios, resolution, frame rate, etc. as will be required with the advent of HDTV. Image compression algorithms exist today that do not suffer from the MPEG affliction, are modular, and are scalable, and achieve this functionality by simple processing such as truncating bits. These algorithms are general in nature, simple to compute, and may provide better video. Some of these algorithms are symmetric. These algorithms will continue to evolve as time proceeds. An open system architecture design will permit decompression algorithms to be downloaded to the set top converter prior to the program material. Different programs could be compressed different ways. Newer programs would benefit from the newer technology, but the hardware at the video resellers and on the customer's set top would remain unchanged! The decompression algorithms would become part of the video material and be stored as a preamble to the compressed video at the video server. Film owners could determine in advance what compression algorithms favored their films. The authors currently have the design for such an open-architecture digital image processing subfunction, and its design is being readied for custom VLSI chip implementation.

Program database contents and selection

Together, the program selection computer, the video server, and the TV set top program selector box permit navigation and control through numerous available programs. Here we look at the available VOD services, information, and search/sort mechanisms that allow the information provider as well as the subscriber to manipulate the database. "VOD" services, as defined in the general trade literature, include but are not limited to such things as: video entertainment programming, video educational programming, interactive games, mixed-media information services, and shopping services. This database will contain information regarding the video content, e.g., program titles, scheduling information, ratings, names of actors and actresses in the program, and textual and/or video data associated with the program content. We will discuss the menuing technology coupled with our single-button point-and-click remote control concept as a vehicle for accessing the complete database of program material.

We envision a system that is intended to facilitate acquisition of program sources by name, content, participants, or some not-necessarily-complete combination of each. For example: "Find Tom Cruise movie released in 1992 related to the law." We conceive a Windows-like menu to appear on the TV screen when and only when the remote control is picked up. The menus would appear across the top of the TV screen (on top of the existing program material). Each of the menus would be pull-down menus. The remote control would have a single-button laser pointer. The subscriber would point the remote control at the menus and be able to pull them down. The remote control would be a full-duplex remote control. It would transmit to the set top box and read the area on the TV screen in the neighborhood around the pointer. This is conceptually how the pointer and pull-down system menu might communicate. Additionally, the remote should possess a conventional learning facility to replace the regular TV and VCR remote controls.

Menu navigation

A typical search might be implemented as follows: The subscriber wants to watch a fight, but does not know when it will occur or who the fighters are; he or she only knows it will occur this month. The customer will point his or her remote to the FIND menu and pull it down. The menu will expand into a number of items such as films, special events, catalogs, information, education, business, news, sports, and weather. He or she could click on either sports or special events, which again would be expanded. Like most other graphic user interfaces to an SQL database, rarely a need exists to click more than 5 times to make the final selection.

The subscriber could click on a VIEW menu to sort program material by program name, actor's name, producer's name, kind of program, and date of program. The subscriber could click on LEARN, which will cause all subsequent requests to be learned; that is, does the subscriber desire emphasis on sports, musicals, new films, etc., or what types of items does he or she prefer when doing interactive TV shopping? The subscriber could click on the COST menu to determine the amount of his or her charges this month, and see a copy of his or her unpaid VOD bill.

The menu information is updated as often as necessary to the subscriber's set top box via the return transmission link from the video retailer. Certain menus can be broadcast to all set top boxes at the same time, or they could be selectively updated. Selective menu information and billing information could be transmitted to individual boxes. A menu called CONTROL would permit the customer to control future events, such as turning on VCR and TV for specific program material or set at a prescribed time and day. The system, via menu clicking, could facilitate other features such as electronic mail and fax retrieval.

Menus and programs would be in full color. Navigating through the menu is analogous to navigating through a Microsoft Windows or Macintosh System 7 environment. This includes help, which would be available at all times.

Program switcher/router

The program switching/routing and selective distribution function is the basic element that differentiates the subscriber loop TELCO application from the other systems. The TELCO subscriber loop facility directs different programs to different copper wire pairs for a space division topology. A unique pair of copper wires brings a single program to the customer premises this way. The other applications use time division and frequency division multiplexed signals to direct programs to their destination. All programs get to all destinations, but selective demultiplexing and decoding bring only one selected program to the TV set.

Program routing is best implemented in a packet switching network. This is a self-routing implementation, as each packet contains the destination, source, and program address. SONET, FDDI, and ATM are all viable implementations, but for TV distribution, ATM may be the preferred method of connecting a subscriber to a program on the server.

ATM, or asynchronous transfer mode, consists of 53-byte fixed-length cells (or packets) that allow insertion of digital signals and their routing information asynchronously. The cell used by a particular subscriber or service remains in a fixed sequence and time domain, although each cell is asynchronous (has no relationship) to all other cells. The ATM cell is 53 bytes long, consisting of a 5-byte header containing the address and a fixed 48-byte information field. ATM is a standard for information transfer developed as part of the overall BISDN effort using the BISDN protocol stack, which has been standardized by the CCITT and ANSI. The ATM switch provides synchronous switching and routing at the speed of the physical medium and the demand rate of the user.

An ATM router provides switching and routing of ATM program data cells as delivered to it by the video server via self-routing modules directed by the address in the ATM cell header. The ATM router provides the electrical interfaces necessary to accept signals from the video server and distribute signals into a network of customers. An ATM switch providing direct connection to telephone interexchange carriers for long distance service via an external ATM to DS-3 conversion multiplexer would permit video wholesalers easy access to video retailers in a pure electronic way. No logistical distribution of magnetic material would be required.

An ATM switcher/router would require:

- 53-byte cell format compliance to CCITT and ANSI specifications
- physical interfaces compliance to DS-1, DS-3, and STS-3c
- video server interface compatibility
- scalable and modular architecture
- fault tolerance
- inclusion of a switch controller and software operating system designed as an open system, based on applicable standards

Modulator/demodulator

The modulation system for a multichannel VOD system could be of a wideband QAM or CAP technology. CAP is suppressed-carrier QAM.

Scalability and modularity

In the case of CATV and TELCO VOD, the system must be modular and scalable so that it is cost effective from 500 subscribers to 100,000 subscribers, and from 20 films or program sources to 3000 program sources. In the case of cellular VOD TV, the system must be modular and scalable so that it is cost effective from 1000 subscribers to 5000 subscribers, and from 20 films or program sources to 200 program sources. In the case of hotel VOD, the system must be modular and scalable so that it is cost effective from 20 subscribers to over 1000 subscribers, and from 20 films or program sources to 200 program sources.

Service management

This section refers to the types of services which should be available on the system, the issues on how to upgrade, and the issues of how to modify services.

- The services that are available include but are not limited to: (1) video on demand/pay-per-view TV, (2) interactive marketing and sales. The diagnostic and maintenance facilities of the VOD system will simplify video retailer's plant maintenance because full loop back diagnostics will exist and operate between the subscriber and the plant, and will operate continuously both day and night.
- The system can accommodate full diagnostics on the full VOD system and the video retailer's network on a continuing basis. Failures will be detected immediately.
- New services can be easily added by programming because of the open architecture and program controlled functions.
- Existing services can be modified by modifying software. Because of the software environment, service modification will correspond directly to specific software changes and/or enhancements. Software modification of the system will modify the service. This will simplify system modifications and reduce their costs.

TELCO Television Applications

This TELCO discussion concerns itself with a standard AT&T Paradyne ADSL technology, which would be evolved into an adaptive ADSL technology that will permit multiple video sources to be transmitted to each subscriber.

Asymmetric digital subscriber line (ADSL) technology permits transmission of 1.544 Mbps of digital data to telephone subscribers and provides a slower 9600 bps return path. This digital transmission is independent of conventional telephone service, which may also be provided simultaneously on the same local loop. The video on demand (VOD) application uses this capability to transmit a data stream formatted according to one or more standards, such as the moving picture experts group (MPEG) standard, delivering full-motion video and stereo audio to the subscriber's television set. Unlike cable TV, there is a separate transmission path to each subscriber, permitting each subscriber to view a different program. The design should conform to open architecture criteria, which will permit downloading and execution of other image compression and processing algorithms as they become available. This open-architecture philosophy is also applicable to CATV, cellular TV, and hotel systems. As enhanced algorithms become available to enhance the video quality, the system can be upgraded automatically by downloading new algorithms to the set top box. Different programs could use different algorithms, and these algorithms could be downloaded before or with each program. Each algorithm requires only a few seconds of transmission time. Enhanced algorithms could be HDTV compatible and accommodate different resolutions and aspect ratios.

Our TELCO system concept consists of 9 major subsystems (plus TV set) as shown in the block diagram in Figure 3.6. At the central office side, an ADSL splitter separates plain old telephone service (POTS), a bidirectional 9600 bps asynchronous path, and unidirectional 1.544 to 2 Mbps path from the central office (CO) to the customer premises (CP). At the CP side, either a separate ADSL splitter or one integrated into the TV top selector will provide customer program request information to the program selection computer at the CO via the local loop. Because all activity begins with the selector, we will start the discussion there.

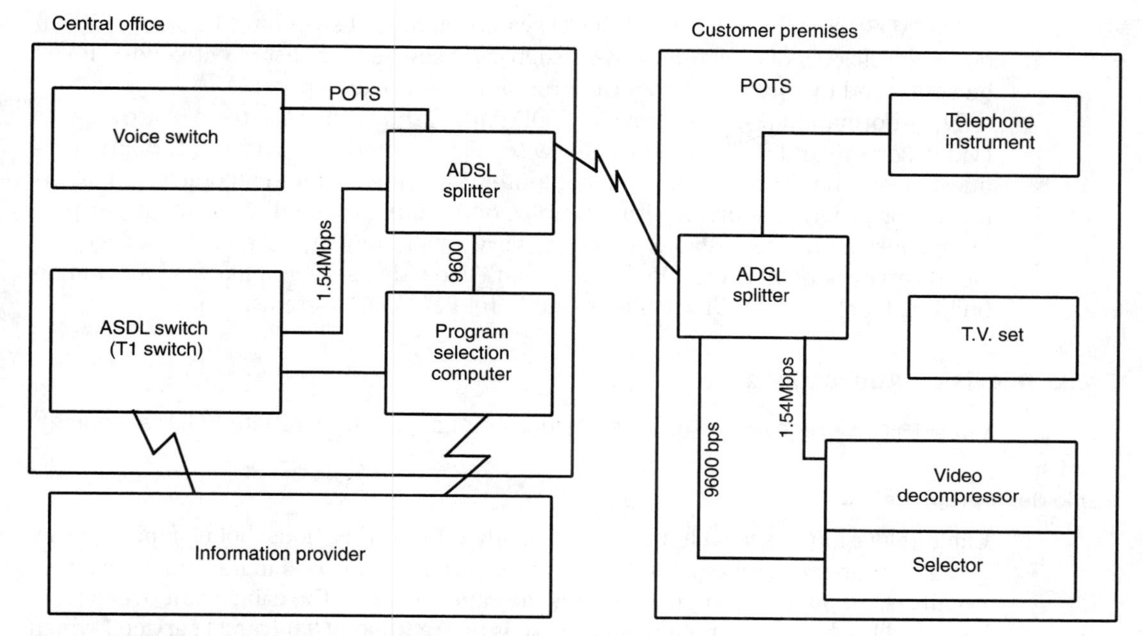

Figure 3.6 TELCO office and subscriber VOD block diagram.

Set top selector at customer premises

The TV set top box extracts the high-band digital signals from the subscriber loop, decodes the compressed video and audio, and presents either channel 3 or 4 modulated or base-band audio and video to the TV set in a format suitable to the TV set (NTSC, PAL, SECAM, HDTV, et al). Customer requests are routed from the selector function in digital format (as a reverse path) back to the TELCO distribution facility.

VOD equipment at the TELCO central office

The TELCO central office must provide support for all TV set top functions, including supplying compressed video and audio, and interactive menus describing program choices, accounting, switch control, etc.

The ADSL *splitter* at the CO separates POTS and digital video service. The POTS path in the CO is not otherwise affected. The ADSL path provides bidirectional digital data from the CO's program selection computer and acquires MPEG (or other) encoded video program sources from the ADSL switch, which is derived from the information provider.

The program selection computer (PSC) communicates with all the selectors of all the customers. It monitors all customer requests and relates that information to the ADSL switch and the information provider. Further, the PSC communicates with the CP selector and provides program and service menus, remote control of TV, VCR, security devices, etc. so that services such as turning devices, programs, etc. on and off can be enabled. The PSC is also a common building block of the CATV, cellular, and hotel PPV VOD systems.

The ADSL switch can be any TELCO switch capable of switching T1 data, or it can be a specialized video switch whose design is maximized for cost effectiveness. It will be controlled by directives from the program selection computer.

The information provider provides VOD data to the retailers' intermediate storage (video server) and then to the ADSL switch. The switch is provided information requests from the program selection computer. The video server can deliver up to "N" movies or program sources with multiples of 5 minutes time displacement per program source. "N" can be a very large or very small number, because the conceived video server is highly scalable and modular. The video server is likewise a common building block of the CATV, cellular, and hotel PPV VOD systems.

Cable Television Applications

Cable television providers are the obvious point at which to initiate VOD technology.

Cable demographics

Cable television is available to approximately 60% of U.S. households, and of these, about 90% are subscribers. Cable subscribers are concentrated in metropolitan areas, because high population density shares the capital costs of the cable plant over a large base. Unlike the telephone industry, there is no tradition of "universal service," which permits averaging of costs over all customers; in addition, TV cable plant is more expensive per mile than telephone plant.

Cable network topology

Until very recently, all cable TV systems were of the "tree and branch" topology shown in Figure 3.7. In this design, the same bundle of radio frequency (RF) signals is fanned out over successively smaller (and cheaper) cables, rather like a water system.

Amplifiers along the cable (about every 1000 feet) make up for transmission losses, but add noise and distortion. Transmission in both directions is possible, but additive noise problems and higher amplifier and maintenance expense discourage it.

Recent designs use a more telephone-like "star" topology, in which the inner portion uses fiber cable to reach a cluster of customers (50–500), as shown in Figure 3.8. The customer premises service entrance is still coaxial cable carrying RF signals, with remote nodes' electronics converting lightwave signals to RF. Because the signal capacity of fiber is considerably greater than that of coax cable, it is convenient to add switching at the fiber-to-coax conversion point to give each customer access to any signal on the fiber, as shown below. Because the RF path is direct and short, 2-way transmission is much more practical than with tree and branch analog networks.

The advantage of the star-and-switch approach is that every signal on the fiber link is available to every subscriber without the need to increase the capacity of the cable (coax, for millions of existing customers) reaching the customer premises. In addition, there may be as many different signals in use by any customer as the coax entrance cable will carry—hundreds of channels, enough to satisfy a multiple-household dwelling like an apartment house. This extremely powerful technique may be used in a hierarchy (e.g., nodes feeding nodes feeding subscribers) to optimize cost vs. channel capacity.

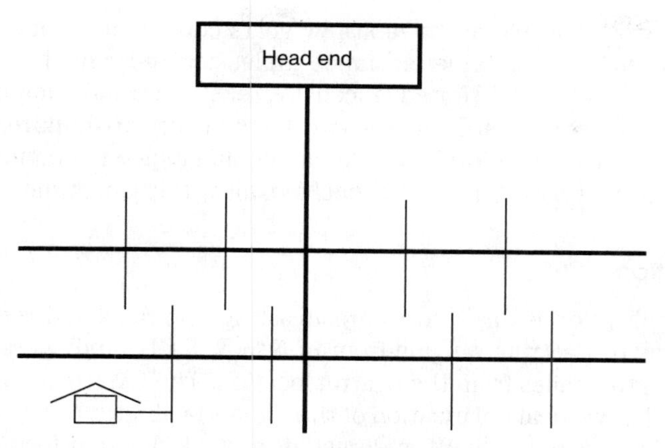

Figure 3.7 Tree and branch topology (all cable coax).

Signals on the high capacity fiber part of the network may be a mixture of analog and digital. Analog TV signals on fiber are easily converted into traditional cable TV channels for use with conventional TV sets, as might be used for "basic" cable service. Digital signals may be multiplexed using any of several techniques, but a packet time division scheme called Asynchronous Transmission Mode (ATM) seems especially promising.

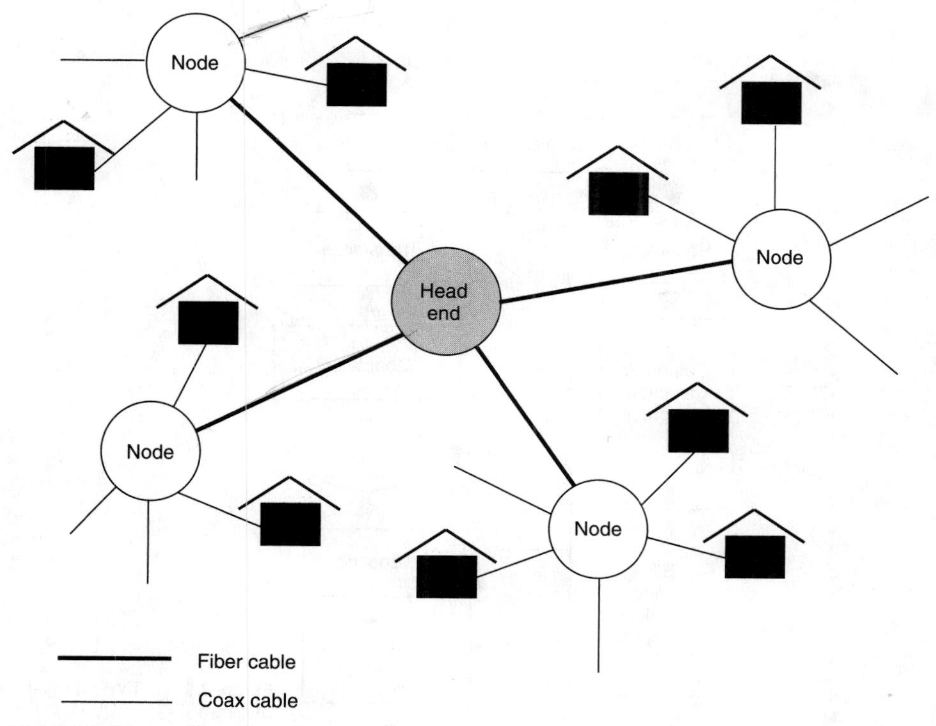

Figure 3.8 Star configuration.

An advanced network feasible in the next few years could consist of fiber-linked nodes connecting to existing customer entrance cables carrying a mixture of analog (for conventional TV) and digital (for advanced TV, telephone, and computers) signals on radio frequency channels. Such a network could support hundreds of television programs, conventional and video telephones, high-speed computers, high-resolution facsimile, and services not yet conceived, all at the same time.

Cellular Television Application

The cellular TV application is directed to broadcasting programs and receiving request data in the radio spectrum ranging from 27.5 to 29.5 GHz and operating short distances of about 4 to 6 miles from the central location. The FM nature and the selection of horizontal or vertical polarization of the signals facilitates easy reuse of the same frequencies in adjacent cells. It is depicted in block diagram form in Figure 3.10.

At the customer premises, a set top box accepts customer requests, transmits requests over the air when polled to the cellular TV service company. The request is received, processed, validated, billed, selected, encoded, and transmitted to the subscriber.

Cellular TV could become part of what has been called local multipoint distribution service by the FCC. The FCC has been asked to set aside 2 GHz of spectrum from 27.5 GHz to 29.5 GHz in each market, for the purpose of deploying this new

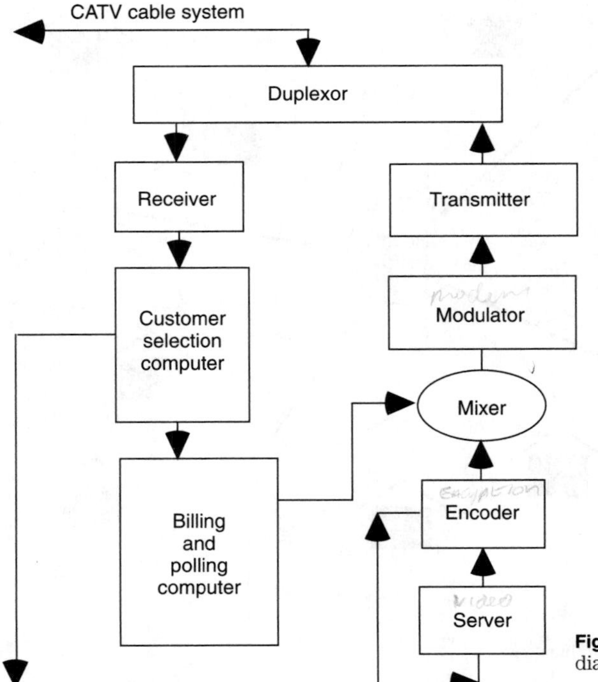

Figure 3.9 CATV VOD system block diagram.

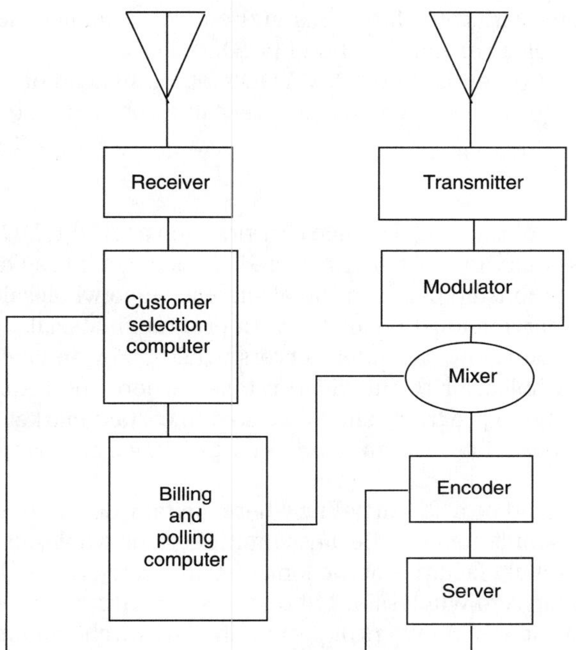

Figure 3.10 Cellular VOD system block diagram.

technology. The 2 GHz of bandwidth would be split evenly between two licensees, thus leaving a full 1 GHz of bandwidth for each licensee to market a profusion of two-way video, audio, and data services.

Using FM technology, the video carriers are spaced 20 MHz apart, and unlike satellite transmission technology, each video carrier within the cell is transmitted on the same polarization.

Propagation through obstructions such as buildings, foliage, hills, rain, etc. is a problem which must be dealt with, because these microwave frequencies are naturally severely attenuated by such obstructions.

An in-home transmitter would be capable of providing a 30-kHz (48 Kbps) data return channel with less than 20 milliwatts of power! These return channels would be frequency interleaved (halfway in-between) among the video signals using orthogonal polarization, in order to ensure effective isolation between the video and data services.

Lodging Industry Applications

Hotel pay-per-view systems can take on a variety of appearances depending on the availability of existing wiring. If a hotel community-style cable is available, the system appearance might assume that of the cellular TV system with the exception of the frequencies used and the fact that the frequencies would not be transmitted into space but into the community antenna cable (less antenna). If it were not economical or not possible to have access to antenna coax distribution system, it might be possible to use the phone wiring to the rooms and employ a miniaturized ADSL/MPEG TELCO type technology around the hotel PABX.

Figure 3.11 represents a potential hotel PPV VOD system. Instead of antennas, a duplexer connects the hotel VOD system to the hotel cable antenna system.

Long Distance Carriers

It may be hard to conceive of the long distance carriers such as AT&T, MCI, SPRINT, etc. as having a role possibility in VOD distribution, but they can if they desire. Figure 3.2 illustrated the relationship between the electronic video wholesaler and the video retailer. The video servers must be loaded with program material periodically. An easy way to load programs into the video servers would be by an electrical connection from the video wholesaler to the video retailer, where the retailer and/or wholesaler decide which video programs should be sold to certain market segments. This connection could use satellite transmission or long distance company transmission facilities.

The nature of the call could be a T1 party-line where digital video is transmitted to many distribution points simultaneously. Because multiple video wholesalers may exist, the long distance switching facilities of the long distance carriers may be of great benefit in connecting Group A to Wholesaler 1 at one time and Group A to Wholesaler 2 at another time, permitting vast numbers of permutations! Switched multipoint dis-

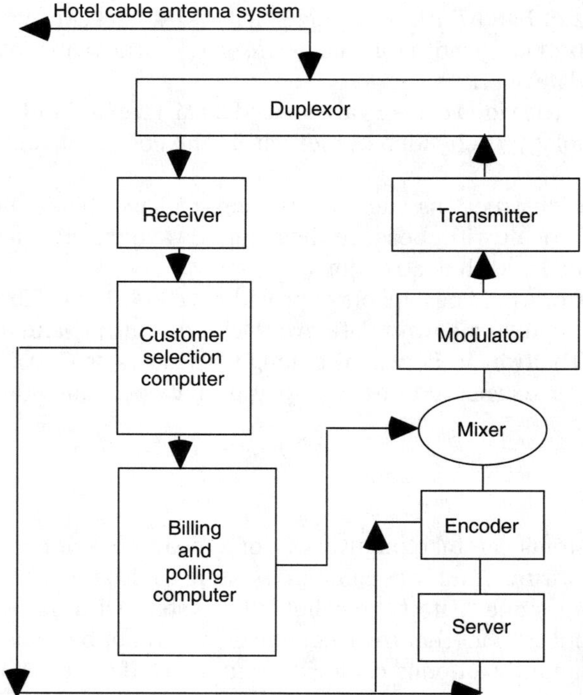

Figure 3.11 Hotel VOD PPV system.

tribution to local VOD retailers could be an added revenue function to the long distance carriers. The fact that the transmission is over copper wire, microwave links, optical fiber, or satellites is totally transparent to the VOD resellers. This would permit the long distance carriers to sell currently unused excess resources.

Summary and Conclusions

The VOD business provides opportunities to a variety of organizations. An isolated start-up VOD retailer might not have all the infrastructure available to him or her, and that does not matter. For example, early loading of the video server may be by magnetic tapes, optical disks, or other mechanisms. As the infrastructure evolves, the pioneering VOD retailers can "patch in" to the new resources as they become available. The VOD business is a business that can start without an established infrastructure, and as the infrastructure develops, attachment to and benefit from it can be derived. Because of the application of modular and scalable systems architecture, and with the advent of video-friendly high-capacity disk drives (such as those available from Micropolis), modern image and video server technology, advanced communications technologies, image compression, encryption, and ATM packet switching, VOD today is becoming practical.

Chapter

4

CATV and TELCO System Network Evolution and Constraints

This chapter provides a brief review of the cable television industry's approach to developing networks and associated infrastructure to facilitate multiple, integrated broadband digital services. The statistical database used to develop this chapter was derived at the Research and Policy Analysis Department of the National Cable Television Association.

The cable television industry passes about 93 million homes, and has about 59 million subscribers in the U.S., of a total of about 94 million homes with television, i.e., 64% of the obtainable market (source: Cable Television Developments, National Cable Television Association, November 1993). Figure 4.1 depicts the growth of CATV subscribers from 1975 to 1993 while Figure 4.2, using the same database, represents the percentage growth each year as compared to the previous year. There are more than 11,000 cable head ends, and the cable TV industry has installed more than 1,000,000 network miles. Installation of fiber continues at a dynamic rate, with fiber plant in the U.S. growing from 24,000 miles in 1992 to more than 41,000 miles in 1993. Construction spending by the cable industry in 1993 was estimated to be about $1.9 billion, of which $1.2 billion was for rebuilds and upgrades. Cable industry revenue from subscriber services in 1993 was estimated to be more than $22 billion, corresponding to an average subscriber rate of about $31 (Cable Television Developments—NCTA). These figures are expected to grow as the cable industry moves into new interactive digital services and telecommunications.

Basic cable revenue rose from $851 million in 1976 to $13,261 million in 1992, while pay TV revenue rose from $65 million to $4.93 billion. Figure 4.3 shows the upward growth of both basic services and pay services. Figure 4.4 shows the relationship of pay TV revenue to basic service revenue. One may hypothesize from Figure 4.4 that VOD, PPV, and ITV revenues as a percentage of total CATV revenue are on a relative decline; however, it must be remembered that: (1) technology necessary

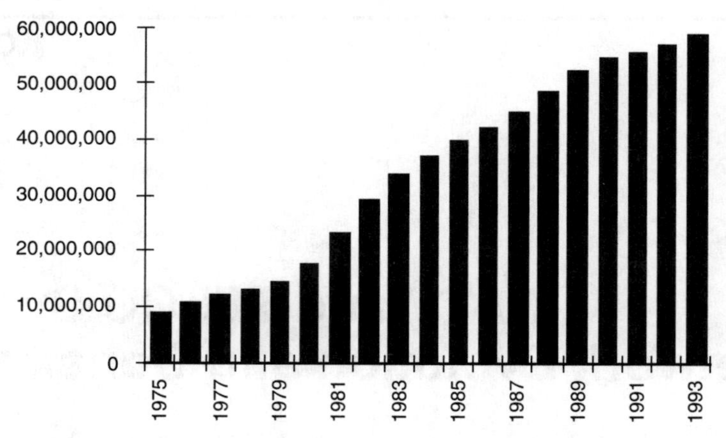

Figure 4.1 Growth rate of CATV subscribers from 1975 to 1993 (data source NCTA).

to support these services is in its infancy and is presently quite limited, and (2) the VCR and video stores started making major impacts in the mid-1980s as providers of in-home entertainment. These numbers do not reflect systems with advanced video servers, ATM switching, digital compression and transmission, and interactive set top boxes. These do not exist (except in experimental form) today, but the need is recognized and they are forthcoming. Pay service revenue is stagnated, because there is no cost-effective technology currently in place to facilitate this next evolution of in-home entertainment, and the similar service could be obtained from the local video rental store with only mild inconvenience. The vastly more convenient pay services are expected to have phenomenal growth when cost-effective technology permits implementation, and this should be in full swing starting 1996. It is expected that after a few years, pay services revenue will exceed basic service revenue.

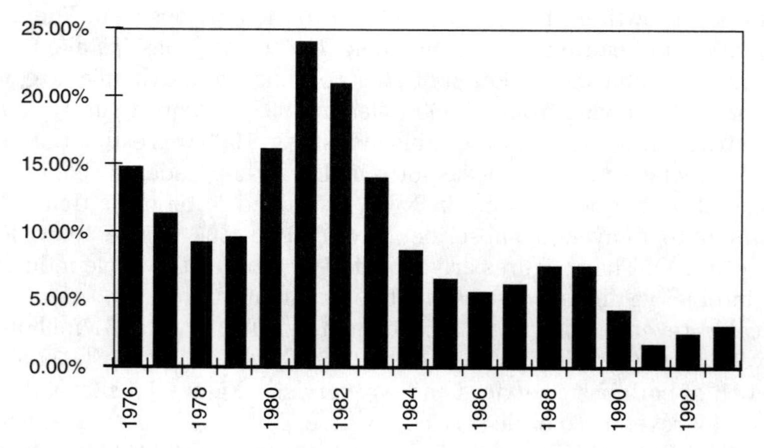

Figure 4.2 Percentage growth rate compared to previous year (data source NCTA).

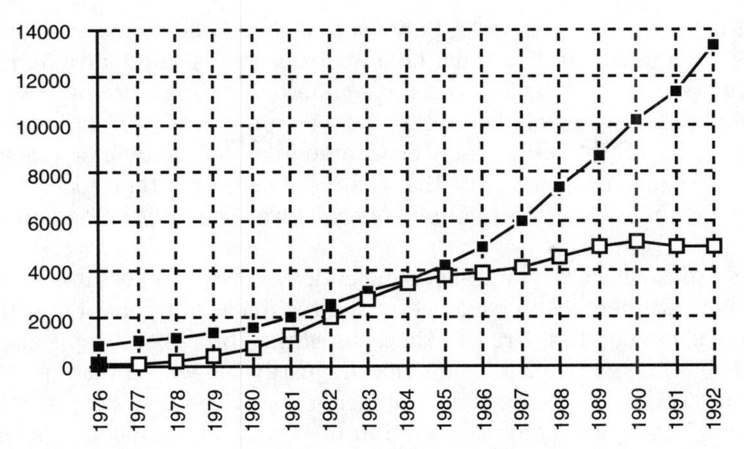

Figure 4.3 CATV revenues in millions, basic services is the dark line and pay services are the light line.

The cable industry's development and deployment of the broadband information superhighway requires architectures for a network infrastructure that integrate the numerous applications that will be offered. Applications will cover a wide range of programs and services. Delivery of video entertainment will evolve from the current core services of the cable industry to enhanced offerings like interactive shopping, near-video-on-demand, and true video-on-demand functions. Telecommunications services will evolve from current voice telephony and data transport to include interactive multimedia applications, information access services, distance learning, remote medical diagnostics and evaluations, computer-supported collaborative work, and more. Many new forms of customer-billable services will be created with this new technology, creating bountiful new pay service revenue streams.

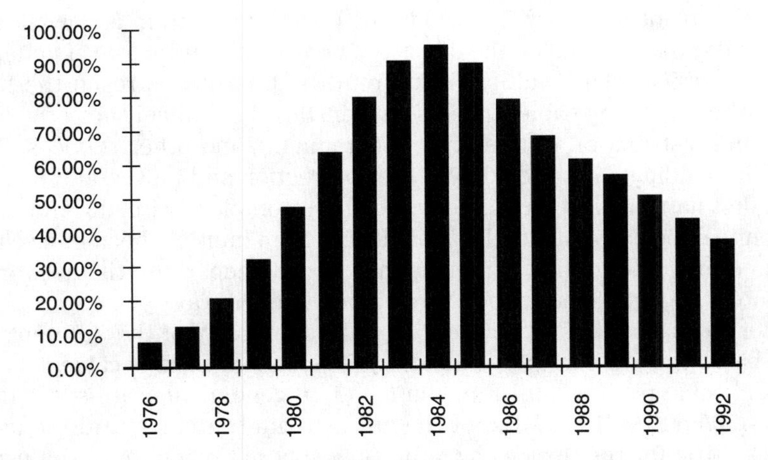

Figure 4.4 Plot illustrating relationship of pay TV revenue versus basic service revenue (database source NCTA).

In addition to the complexity and diversity of the applications, development and deployment of a broadband information infrastructure will combine a number of different networks that will have to work in a coherent manner. Not only will users be connected to different regional networks, but the sources of information will also belong to different enterprises and may be located in remote networks. It is important, therefore, to realize from the start that the two most important goals of the architecture for the broadband information superhighway are integration and interoperability.

Achieving these goals is a mind-boggling endeavor. Nonetheless it is a critical goal that goes beyond merely the elegance of the technical solutions. First, the investments that will be required to realize these networked broadband applications will be quite large. It will be difficult to economically justify the widespread deployment of disjointed networks. The value of communication services grows exponentially with the number of interconnected users, and future broadband applications will have to reach globally to large numbers of customers in order to be commercially viable.

Second, the multiplicity of broadband applications that one can envision in the future will not emerge from a centralized planning and deployment process, but rather from an information marketplace where independent application developers will compete for the customer's attention. The existence of integrated and interoperable networks will stimulate the rapid development of new applications and hence speed the introduction of new services by providing users with added value.

The cable industry's broadband integrated services network architecture should be based on a hierarchical deployment of network elements interconnected by broadband fiber-optic and coaxial cable links. Let us skip ahead a few figures and examine Figure 4.8 to get an overview of this architecture: Starting at the home, a coaxial cable tree-and branch plant provides broadband two-way access to the network. The local-access coaxial cable plant is aggregated at a fiber node, which marks the point in the network where fiber optics becomes the broadband transmission medium.

Current expectations are that approximately 500 homes will be passed by the coaxial cable plant for every fiber node, with variations (from as low as 100 to as many as 3000) that depend on the density of homes and the degree of penetration of broadband services. The multiple links from the fiber nodes reach the head end, which is where existing cable systems have installed equipment for origination, reception, and distribution of television programming and other services. The head ends are in buildings that provide weather protection and power, and hence represent the first natural place in the network where complex switching and processing equipment can be conveniently located. Traffic from multiple head ends can be further routed over fiber optics to regional hub nodes deeper into the network, where capital-intensive functions can be shared in an efficient way.

In order to achieve the integration and interoperability of the different services, the first basic issue to consider is bandwidth allocation. Today's cable networks use the coaxial cable plant as an RF medium and, in general, allocate spectrum as follows: 50–450/550/750/1000 MHz for the downstream traffic (toward the users) and 5–30/35/42 MHz for the upstream traffic (low-split approach, from the users). See chapter 10, Communications.

Interoperability is expected to be facilitated by the use of industry standard technologies, in particular synchronous optical network (SONET) and asynchronous transfer mode (ATM) multiplexing and switching elements. It appears that ATM will eventually become the prevalent solution all the way to the user interface at the home or office, but alternatives are being explored for the early deployment of broadband services in a cost-effective manner. Other mechanisms for internetworking may have to be used until competitively priced ATM products, including interfaces, multiplexers, and switches, become widely available over a range of speeds from a few Mbps to Gbps.

Issues that still need to be resolved in ATM are numerous. They include access and congestion control, quality of service definitions, choice of adaptation layer standards, and transport of compressed-packet video (MPEG in particular). Consensus on end-to-end latency with regard to isochronous services (voice communications in particular), advanced signaling protocols and connection control systems, and management information models and interfaces are also pending resolution.

The final evolution to an integrated services broadband network will require consideration of many additional interfaces and protocols associated with the many services and network interconnection points that will be supported. Signaling protocols, directory access services, and application development environments are only some of the issues to be resolved.

Integrated network management, operation support, and business support systems are important components of the broadband architecture. The use of open and modular platforms and protocols for the development of support systems is a crucial factor in controlling the complexity of the multitude of network components. New applications will be created for the network operators. The evolution will be toward the use of distributed computing environments, coupled with modern software technologies such as managed object models, client-server architectures, and contract trading capabilities.

Cable television plant is evolving from a traditional coaxial tree-and-branch architecture to a ring and bus architecture, in response to technological opportunities and evolving service demands.

Coaxial Networks

It is important to discuss the previous generation of cable architecture, because this describes a great deal of the installed base. It has almost entirely been superseded as an architecture for new builds and rebuilds by the fiber-to-the-serving-area (FSA) architecture that is described later in this chapter. The basic components of the coaxial plant of a cable television system are shown in Figure 4.5.

The basic system components are as follows:

Head end equipment. The head end equipment is the location where television program source material is collected by direct reception, via satellite, from local sources directly or by upstream contribution links, and by various means of storage. This source material and other signals are assembled in frequency-division multiplex (FDM) in an RF band, and this signal is launched through the trunk and feeder sys-

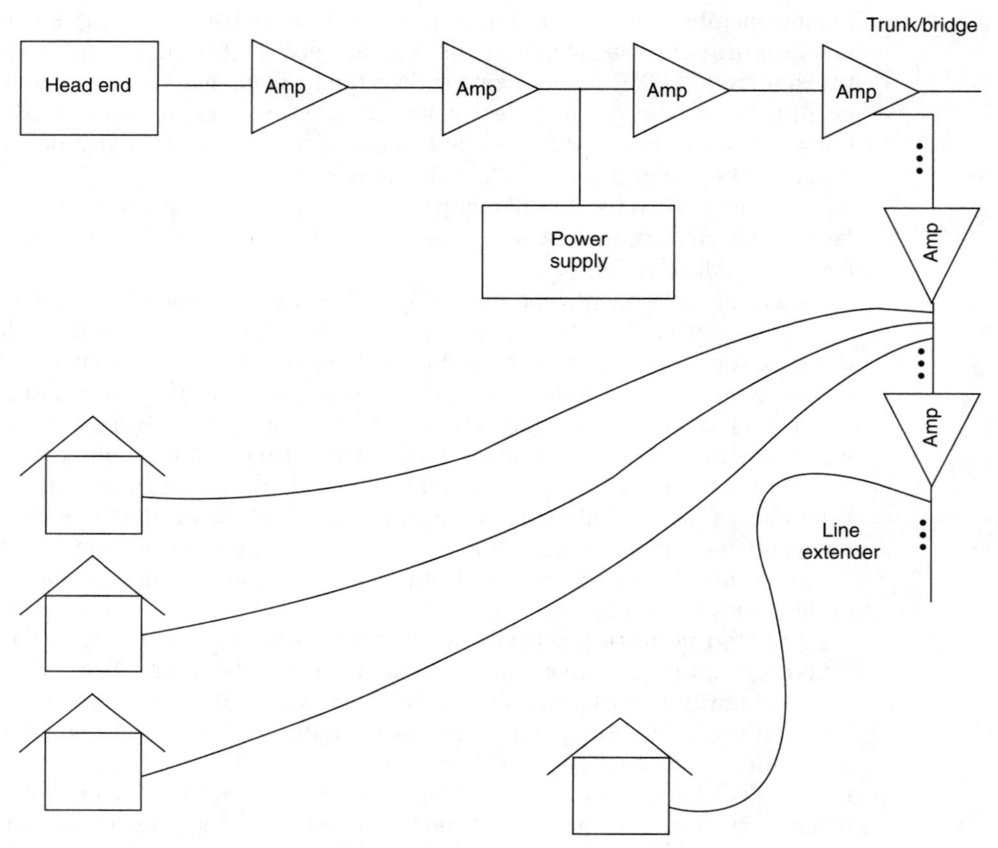

Figure 4.5 Basic components of coaxial plant.

tems "downstream" toward customers. Head ends may serve a few hundreds or thousands of subscribers in small localities, and tens or hundreds of thousands in larger localities.

Trunk cables. Trunk cables are coaxial cables that carry the RF signal from the head end to groups of subscribers. Amplitude-modulated, wide-band microwave systems are sometimes used to bridge obstacles or large distances. Losses in the trunk coaxial cable are made up by trunk amplifiers, which are designed to balance degradations due to low-level noise and distortion. Fiber-to-the-serving-area (FSA) designs now limit the number of active amplifiers after a fiber node to a range of four to ten.

The feeder cables. Feeder cables are coaxial cables that run down the streets or rear easements in the served area and which carry the RF signal from a bridge amplifier on the trunk cable. Bridges might be stand-alone units, or might be combined in a bridge amplifier. Taps, which are directional couplers, are placed periodically along and in series with the feeder cable in order to provide service connection points for drops. Losses in the feeder system are made up by line extenders, which are designed for gain rather than for low noise. Typically there are at most 2 line extenders in any feeder.

Drop cables. Drop cables are coaxial cables that connect residences or other service locations from a tap on the nearest feeder cable.

Powering

Power supplies for the trunk, bridge, and line extender amplifiers are typically placed at suitable locations along the coaxial cable system. The power supplies acquire 60-Hz electric power from the commercial grid, converted to 60-V 60-Hz quasi-square-wave power, and impressed between the coaxial cable center conductor and shield. Each amplifier passes this power through, while taking enough to power itself. Several amplifiers are typically supplied by one power supply.

Standby power

Unlike TELCO implementations, CATV standby power is typically not provided in cable systems that are designed for the delivery of entertainment to homes, because the commercial power grid also supplies the subscriber viewing device. Some systems, however, include lead-acid batteries in their power supplies to compensate for brief localized power outages that could affect a power supply, but not all of the homes are covered by it. These systems provide from 30 minutes to 2 hours of standby power. Extended power outages of limited geographic scope are covered by portable generators. Some head ends are provided with standby generators.

Status monitoring

Surveillance is typically not provided in cable systems that are designed for the delivery of entertainment to homes, although the policy of some cable operators is to provide remote status monitoring of trunk and bridge amplifiers.

Express feeder

To provide better control over losses, and to allow power to be removed from the taps because it can cause connector corrosion, some systems use an express feeder system as in Figure 4.6. There are no taps in the cables that feed the line extenders, and all the taps are fed from separate cables that feed forward and backward from the nearest amplifier.

Bandwidth

The lower edge of the downstream band is usually 50/54 MHz. The most common upper limits are 300, 330, 400, 450, and 550 MHz. Some systems are being constructed or rebuilt for 750 MHz; sometimes the spacings between gain blocks are set for 750 MHz, but the amplifiers are 550 MHz. A very few systems have been built or are under construction with an upper limit of 1000 MHz. If used, the return path from the

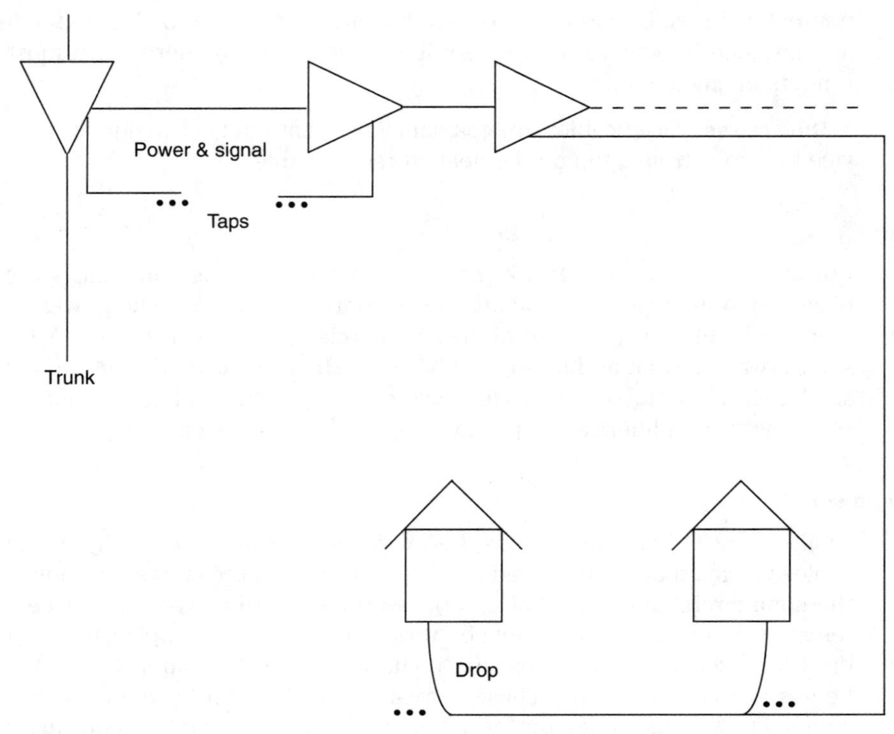

Figure 4.6 Express feeder.

home is typically based on the low-split approach using the 5–30/35/42 MHz band, but other approaches are being tested, such as using the high frequency 900–1000 MHz spectrum (high-split approach) or installing additional coaxial cable (shadow cable approach).

Fiber to the serving area

The architecture of choice for current new construction and rebuilds is in Figure 4.7. The major advance is the substitution of fiber for the coaxial trunks. Typically, at the head end the downstream FDM signal is separated into blocks of up to 80 television signals, and each block amplitude-modulates a laser operating at the 1310-nanometer wavelength. Each laser in turn is connected to a single-mode fiber.

The fibers are connected to "fiber nodes" that are located in the vicinity of 500 to 3000 homes. At the fiber node, photodiodes recover the block of television channels from each fiber; the blocks are then reassembled to form a single FDM RF signal, and the RF signal is applied to coaxial trunk and feeder plant and is delivered to customers.

This type of construction is frequently less costly than the all-coaxial equivalent. It provides significant improvements in transmission quality, greatly increased bandwidth, and is less demanding of maintenance and repair resources.

Conditional access

It is typical for various packages of services to be offered to customers for different prices, and the service available to individual subscribers is configured accordingly. Band-reject and bandpass traps are often used to perform this configuration. In addition, active devices are provided in a set top box or in the cable plant to allow a customer to receive one or more specific channels for specified periods of time. These are conditional-access devices. The controls for conditional-access devices may be at the head end, at a central location of the cable operator, or they may be national. This forms a specialized "intelligent network."

Several proprietary systems are in use. In Canada, for example, the Canadian Radio-Television and Telecommunications Commission (CRTC) has recently encouraged the development of a single national standard for conditional access.

Digital video compression

Television services are now delivered in National Television Systems Committee (NTSC) format. This analog format, in which the picture is amplitude-modulated ves-

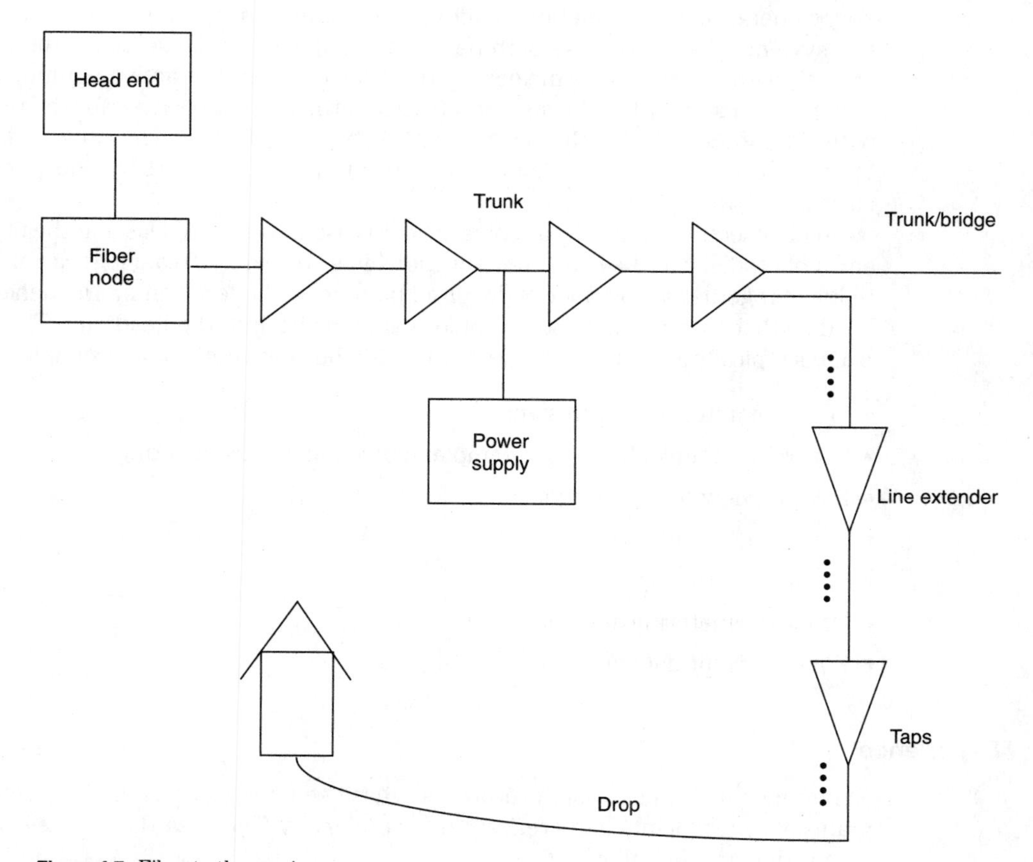

Figure 4.7 Fiber to the serving area.

tigial side-band and the sound is frequency-modulated, occupies 6 MHz per television signal, and the FDM RF signal assembled at the head end is fundamentally based on 6-MHz channels.

Some cable operators have placed orders for large volumes of digital video compression (DVC) set top boxes, which will decode digitally compressed television and other signals. Typically, compressed program sources will be delivered to head ends via satellite or recordings, and will be transmitted to subscribers as a multiplexed digital signal within one or more 6-MHz channels. Various classes of service can be accommodated with different bit rates, including high-definition television. The set top boxes that have been ordered will decode one digital signal at a time, and will transform it into NTSC at baseband, or RF for connection to the customer's TV set. Conditional access can be integrated into the set top box.

Some manufacturers of set top boxes have indicated a readiness to accommodate other services, including telecommunications services, in their systems.

Upstream, or reverse path

Current practice is a low-split approach. Most systems are capable of being set up for 2-way operation by installation of additional modules in the amplifiers, but in only a few systems has the reverse path been activated. On coaxial sections, the low-split reverse path is created by means of crossover filters in the active equipment. The high pass is used for the downstream path and the low pass is used for the upstream path. The lower edge of the reverse path is typically 5 MHz. The upper edge is 30 MHz in most systems, 33 MHz in some systems, and 40 or 42 MHz is now being discussed with manufacturers.

On fiber sections, typically a separate fiber is used, with an AM laser at the fiber node and a photodiode at the head end. The individual coaxial upstream legs can all be multiplexed together at the fiber node, or all but one of the legs can be frequency-translated so that they all have individual logical paths back to the head end. The reverse plant is typically activated only where there is a business case for it. Examples include:

- Status monitoring return signals
- Contribution links from, for example, municipal council meetings
- Pay-per-view ordering signals
- Traffic-light control circuits
- Security alarm signals
- Other telemetering and control circuits
- Point-to-point data and telephony signals

Experience

Experience with the low-split return path has been mixed, and currently active programs of return plant characterization are underway. The hazards and their probable disposition are the following:

Ingress. The coaxial plant is in an environment of high ambient radio signals, mainly from short-wave broadcasts around the world but also to a lesser extent from citizen's band, amateur radio, and other legitimate users of the high-frequency spectrum. Under conditions where the shielding effectiveness of the cable plant is compromised, these signals can enter, or ingress. Points upstream in the cable plant experience a funneling effect of the summation of all ingress points. The effect can be harmful to upstream traffic even if the cable plant is meeting regulated limits for downstream leakage into the air. The effect by frequency is diurnal, depending on behavior of the D, E, and F layers in the atmosphere, and also depending on the sunspot cycle. The parts of the plant closest to the customer are also the worst contributors, with TV sets and other customer equipment contributing worst.

Possible means of combating low-split ingress include:

1. Making the fiber node sizes smaller
2. Filtering off those legs and drops where upstream signals do not need to be supported
3. Ensuring tight joints in the cable system; regeneration of digital signals at suitable points
4. Using modulation schemes such as spread spectrum that can combat the ingress
5. Using narrow-band modulation schemes that can fit between the major blocks of ingress (e.g., between the international short-wave broadcast bands)
6. Using adaptive modulation schemes that can change frequencies in response to a change in ingress level will also be of help

Impulse noise. Impulse noise is present in both directions of transmission. It may arise from points of physical weakness or looseness in the plant, and may also be due to induction from external sources such as power lines. Impulse noise generated from electric lines is more prevalent at lower frequencies. Cable systems seldom have critical impulse noise problems in the downstream distribution plant because cable shielding for frequencies above 54 MHz is generally sufficient to keep impulse noise at a low level.

Possible means of combating impulse noise include: using digital transmission schemes with forward error correction, making fiber node sizes smaller, filtering off those legs and drops where upstream signals do not need to be supported, ensuring tight joints in the cable system, and using cable bonding and grounding.

Bandwidth for new services

The bandwidth requirements of new services, such as personal communications services, video on demand, voice telephony, video telephony and high-speed data, will depend on the technology used and the traffic generated by customers.

If the low-split 5- to 42-MHz band will suffice for the traffic from any customer, then it will be a design issue as to how close to the customers the fiber nodes have to be located in order to provide sufficient upstream bandwidth for all of the offered services. It can be assumed that, once the return signals reach a fiber node, bandwidth will not be a limiting factor.

Some other alternatives that have been considered are:

1. High-split return, in which the return path is above the downstream path. Testing is underway on the use of the high-frequency spectrum (900–1000 MHz) for return transport. Preliminary results indicate that the high-split approach represents a potentially cost-effective solution to having high bandwidth in the return path. However, there are still technical challenges that need to be overcome before this approach can be widely deployed.
2. Shadow cable return, in which the return path is provided by a separate coaxial feeder cable equipped with 1-GHz passives as needed. The associated RG6 drops to the premises are installed on activation of the service. The shadow coax is operated passively for telecommunication services, and although passive, the anticipated bandwidth allocation scheme is 5–175 MHz upstream and above 225 MHz downstream.
3. Mid-split return, in which all television customers are provided converter boxes, and the band normally occupied by TV channels 2 through 6 is allocated to the reverse path. It is not clear whether there would be customer concerns about the widespread deployment of converter boxes, and regulatory issues might preclude it.
4. While the initial provisioning of two-way services will rely on the available low-split band, the use of upstream bandwidth in other ways (e.g., high-split or shadow cable) represents potentially attractive long-term options.

Regional Hub/Passive Coaxial Network Architecture

Cable plant evolution is moving toward a regional hub/passive coaxial network architecture. Some of these changes are already being deployed by cable operators. The main changes from the fiber node architecture are summarized below.

Regional hub

Frequently in a metropolitan area or region will be found a multitude of head ends, each serving perhaps a few tens of thousands of homes, but which are replications of each other. By joining them together on a fiber ring, and placing on the ring a superior hierarchical location called a regional hub as shown in Figure 4.8, some of the functionality of the head ends can be more economically realized at the regional hub. This is particularly true for services where the technological building blocks are expensive.

Examples of cases where functionality could be more economically deployed at a regional hub could include earth stations, commercial insertion, video-on-demand platforms, multimedia platforms, and telephone switches.

Regional ring

Some regional rings have already been deployed in which television programs are carried on AM, FM, and proprietary digital fiber systems.

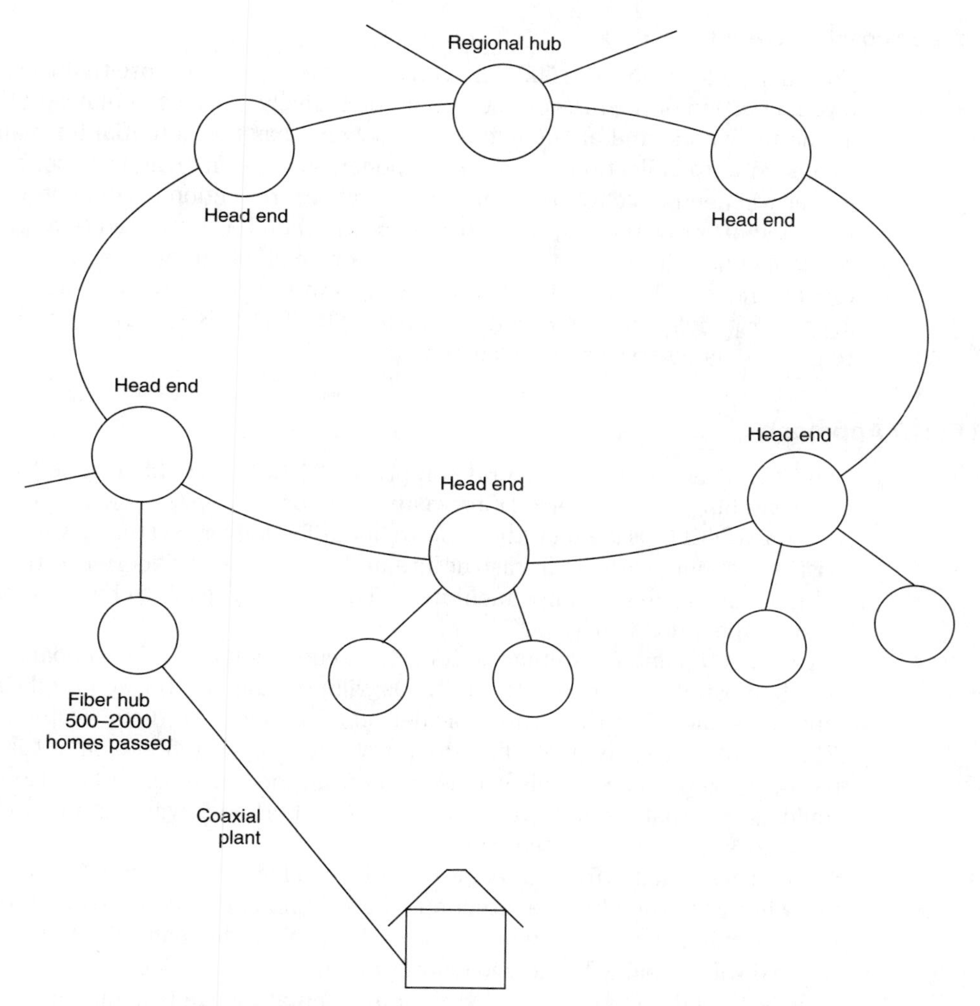

Figure 4.8 Regional hub fiber/coax network architecture.

Under active discussion is the potential for future deployment of asynchronous transfer mode (ATM) and distributed-queue dual-bus (DQDB) technology.

Fiber hub

It is currently economical to deploy fiber nodes at the level of about 500 homes passed. Concurrent with this evolution, it is expected that it will be appropriate to introduce an intermediate hierarchical level, called the fiber hub, between the head end and the fiber nodes. This is shown in Figure 4.8. The fiber hub will perform a fan-out function and will also be a convenient location for alternate routing (it is not foreseen at present that alternate routing will be required below the fiber hub level). Transmission arrangements for the head end-fiber hub section are under discussion.

Passive coaxial network

For fiber nodes below the 500 homes-passed level, some propose that the remaining coaxial portion of the network be operated passively. This may imply additional gain at the fiber node and at the home. This scheme has the potential for maintenance savings due to collection of active components at the fiber node. Also, because no power will need to go to the feeder, there will be a reduction in connector corrosion. More importantly, though, the subscribers will then be connected to a passive bidirectional bus, with its significant potential for flexibility in new services, bandwidth on demand, etc. The passive bidirectional bus would enable the use of shared-media access protocols (such as variations of IEEE 802.6, DQDB, CSMA/CD, and Ethernet) to provide data services to the customers.

TELCO Approach

The TELCO approach is a hybrid comprising TELCO-style fiber-optic transmission and switching, and either ADSL/twisted pair or CATV-style coaxial cable to the home. The TELCOs, being rather new to home TV delivery systems, will look to their existing communications infrastructure and to the CATV experience to determine the most cost-effective implementation. The final approach to be selected will depend on the individual TELCO.

An expected and continuing trend is the acquisition of CATV companies by TELCOs. In most of these cases, the TELCOs will probably use the acquired CATV company's installed infrastructure for delivery of video to their customers. Some TELCOs are currently upgrading their systems to optical fibers, and it is expected that newly deployed systems will use coax from the hub to the home. However, one would expect that a significant number of neighborhoods will continue to be served by the existing ADSL/twisted pairs.

California's Pacific Telesys seems to prefer the hybrid approach. Of course, the future will clarify the ultimate direction taken. Although it is obvious that a single approach would provide economies of scale, because of the many TELCOs nationwide, this goal will probably be slow in coming about.

As described previously, the hybrid implementation can be a mix of:

- TELCO-only technology
- TELCO and CATV technology, or
- CATV-only technology

The TELCOs' final choice of implementation should have its basis in cost, quality, capacity, viewer responsiveness, maintainability, reliability, and flexibility. Figure 4.9 represents a hybrid TELCO application. It should be noted that deleting either the twisted pair or the CATV-style coax delivery system would make this hybrid system a homogeneous system with either a CATV or TELCO flavor.

The TELCO head ends can be connected in a hub configuration just as the CATV systems shown in Figure 4.8. Alternatively, it could be a stand-alone head end, or a group of head ends communicating by existing TELCO DS "x" switching and transmission protocols.

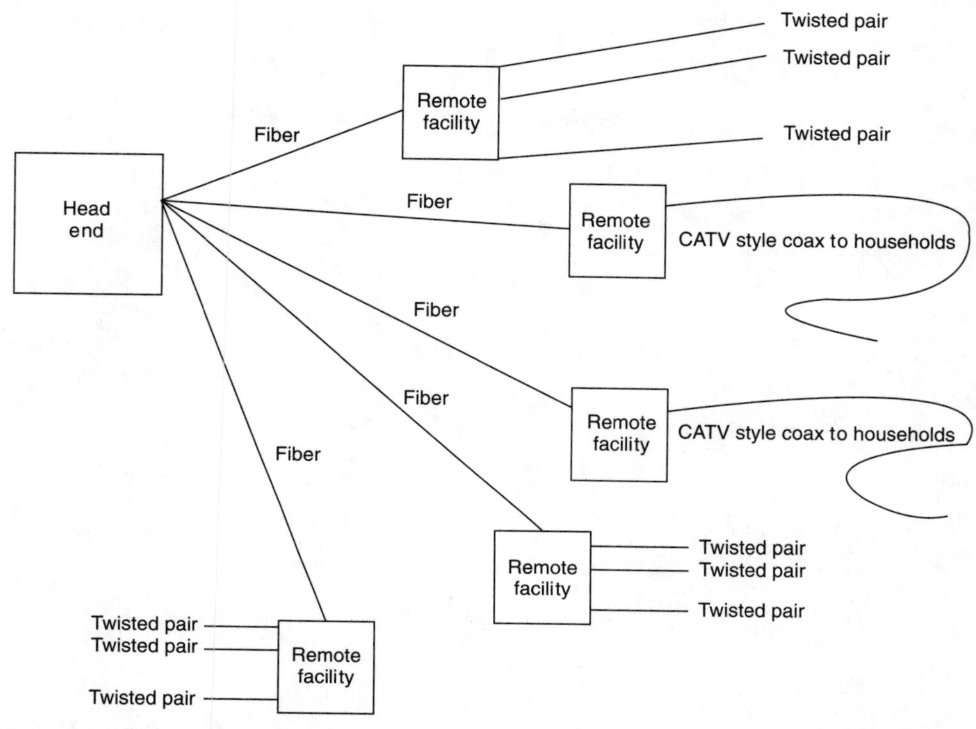

Figure 4.9 Hybrid TELCO system using TELCO and CATV components.

Conclusion

Video on demand is the next great land rush. Telephone companies are currently restricted from offering broadcast services such as the CATV companies do. VOD has legally been defined as transactional and can permit telephone companies the opportunity to take significant market share in delivering television programming over "twisted pair" phone lines (or coax) before CATV companies try to take market share in delivering telephone calls over their cable facilities. Will the telephone companies, who have nearly unlimited access to capital and thus have the ability to purchase spectacular amounts of equipment, force the cable companies to raise capital and purchase equipment to defend their current market from erosion? Will there be an ITV war? While the approach and winner may be uncertain, one thing will be certain: The TELCOs will raise capital and purchase equipment to defend their current telephone market and be a viable VOD contender.

Chapter

5

Image Compression, Cost, Quality, Technology, and Philosophy

This chapter will not attempt to derive a large number of intricate formulas and equations, nor will it reference the numerous works of others on the subjects of advanced communications theory, probabilities, statistics, and linear operators. Rather, this chapter demonstrates the need for image compression and will summarize some salient features of some of the more important image compression algorithms, with an attempt to conclude that it may be shortsighted to select only one algorithm and legislate its exclusive use. I will attempt to illustrate: (1) why media creators should have control over their entire artistic contribution, including how it will be preserved and transmitted, and (2) the undesirable consequences of visual and audio artifacts various compression algorithms may produce. What will the image look like? How will the program sound? This chapter will deduce that perhaps final image coding methodology should be the responsibility of the major motion picture studios, because the motion picture studios are the ultimate guardians of the media quality.

Ever since the advent of modern digital computing in 1962, ever since digital simulation was born, ever since satellites sent myriads of image data from space, there has been a need for image compression. Many believe digital image compression was born in the 1980s, when it was used or popularized in a variety of telecommunications applications, namely teleconferencing, digital broadcast CODECs, and video telephony; but that notion is incorrect. In the early sixties, the Earth Resource Satellite Technology project was born, which was commonly called ERTS and popularized in many journals, including *National Geographic*. ERTS was the forerunner of Landsat. These satellites orbited the earth scanning the planet's surface through their multispectral sensors, whose vision ranged from ultraviolet to infrared using many parallel scanning sensors.

Periodically, the entire surface of the earth was mapped using these multispectral sensors, each producing monochrome images in their own spectral interval. Combined, they produced real or pseudo-color, and when processed in combination would yield information about the earth's surface and subsurface. Digital signal processing produced data useful in determining hidden mineral or oil deposits, water quality and quantity, urbanization, military complexes, etc. This information was transmitted from orbit, processed, compressed, and stored in an image database back in the 1960s. Image processing and compression technology is, therefore, over 30 years old.

In those days, each monochrome image sensor was digitized to produce 8 bits of information. Eight sensors would produce 64 bits of information per pixel. Terabytes of data were collected, compressed, transmitted, and stored.

In the 1980s, videophone prototypes were built and field trials were implemented, and the results of these activities were validated. A challenge existed to create full-motion color video telephone service that operated over the plain old dial-up telephone system (POTS), *not* using ISDN nor any other TELCO facility other than their conventional limited-bandwidth dial-up lines. These attempts quite often had partial success. Some of these systems used transform technology such as FFT and DCT to assist in the image compression process. Some systems differentiated between background and foreground imagery, placing emphasis on the foreground and initially ignoring the background until data capacity permitted painting the background. Virtually all the systems relied on removing not only spatially redundant data but also temporal or frame-to-frame redundancy. Image compression, therefore, is a subject abundant with mathematics-based theories, complicated algorithms, and advanced technology, with objectives strongly implanted in difficult-to-quantify human psychological considerations.

Requirements for Image Compression

The MPEG specification makes provision for an NTSC screen of up to 720 by 480 pixels (345,600 total pixels) or a PAL system with up to 720 by 576 pixels (414,720 total pixels). Figure 2.5 (from chapter 2) represents the various number of pixels used in various computer and television systems, and chapter 2 explains how those numbers are derived. Using these NTSC numbers in this chapter and assuming 24-bit color, an NTSC system could require 345,600 24-bit pixels, or 4,147,200 bits per frame for representation. Thirty frames per second therefore require 4,147,200 times 30, or 124,416,000 bits per second. Assuming a broadcast-quality 7-Mbps channel data rate were used, this would require an image compression factor of almost 18:1.

Viewed another way, compared to analog TV using quadrature amplitude modulation and using 4 bits per hertz, the uncompressed digital channel would require 31 MHz of bandwidth rather than 6 MHz. This could be considered only a 6:1 compression requirement. Irrespective of the viewpoint, one thing is evident: compression is required.

Some people believe that a single standard for video compression techniques should be mandated for cost reduction and interoperability purposes. This might be true if it were not possible to build a general-purpose digital image processor (DIP) that was essentially algorithm-independent, but the world often looks toward dedicated pipelined applications that are algorithm-specific. Digital signal processing (DSP) elements are intended for one-dimensional signal processing solutions and therefore are not well-suited to real-time, cost-effective image processing. Images are 2-dimensional functions in length and width. A DSP is a good 1-dimensional processor, but not a good 2-dimensional processor. A DIP designed for good 2-dimensional processing makes possible a general-purpose, real-time image decompression facility that is not algorithm-dependent. This permits an open image-processing architecture facilitating the use of different algorithms with different video programs. A motion picture or TV program can have its own specific decompression instructions, which are attached to the beginning of the program to be downloaded to the image decompressor as firmware.

All aggressive image-compression technologies remove more than redundant information; they remove information that is determined *a priori* not to represent a significant degradation of image quality. This is a judgment call on what is acceptable imagery. Image complexity would normally be related to channel bandwidth, but image complexity varies from image to image. Information channels have finite capacity. When image information exceeds certain complexity, all image compression algorithms determine what excess information to throw away. When images are reconstructed, the reconstructed missing information can be perceived as undesirable artifacts in the image. Algorithms that produce variable-rate data bit streams to preserve image integrity are desirable if variable-rate packet communications facilities such as ATM (see chapter 10) are available. While it is true that an ATM system has a finite total bandwidth, that bandwidth can be shared between multiple uncorrelated increased-quality programming sources, so that increased bandwidth in one program may be accommodated by bandwidth reductions in one or more other programs. When 20 or so program sources share one ATM channel, it is highly unlikely that the total available bandwidth requirements will ever be exceeded. In the unlikely event that an attempt to exceed the maximum total channel bandwidth was made, image degradation would occur.

Many industry professionals believe that the artistic content of the video medium should not be altered by persons other than the media creators. The artist created the mood, not to be disturbed or altered by an impersonal algorithm selected by an uninvolved technician. Different algorithms produce different artifacts, which can include splotches, speckles, blurs, mosaics, lines, blocks, incorrect shading, colors and hues, missing frames, etc. If it were deemed desirable, it would be possible, for example, to perform a background/foreground separation and provide different compression factors to each, thereby preventing serious consequences of overtaxing the channel, and hence creating uncalled-for artifacts. For example, the foreground could be painted in high resolution while the background could be painted in low resolution. The background could be low-pass filtered and then combined with the foreground. The effect would appear as if a very large aperture was used on the camera at the time of the filming.

Of course, this might also be unacceptable media alteration. If the choice of compression algorithm was given to the artist at the time the film was being digitized, any visible artifacts would be under judgment and control of the artist. It is conceivable that different scenes could use different compression algorithms within the same movie, and that these algorithms could be downloaded to the set top box as firmware decompression instructions for the DIP on a scene-by-scene basis. The artist could more carefully control the mood of his medium.

Numerous still-image compression algorithms abound, both transform- and nontransform-based. The better-known transform procedures include:

- adaptive DCT
- Berlekamp-Massey algorithm
- discrete cosine transform (DCT)
- discrete sine transform
- Fano algorithm
- Fast Fourier transform (FFT)
- Fourier transform
- Goertzel algorithm
- Haar transform
- Haar-slant transform
- Hadamard transform
- Heartley transform
- Hough transform
- Karhunen-Love transform
- Laplace transform
- Rademacher functions
- Slant transform
- Viterbi algorithm
- Walsh transform
- wavelet

and nontransform procedures include:

- adaptive DPCM
- adaptive run length coding
- arithmetic coding
- block truncation coding
- differential pulse code modulation (DPCM)
- fractal coding
- multicomponent coding

- polyhedral encoding
- progressive coding
- pyramidal coding
- recursive binary nesting
- run length coding
- sub-band coding
- vector quantization

It can be seen from the few still-frame 2-D image compression algorithms represented that there is no scarcity of image-compression algorithms. And image-compression algorithms continue to be invented.

For full-motion images, in addition to the above algorithms, we have:

- conditional replenishment coding
- predictive coding
- motion compensation coding

The choice of a coding scheme might be precarious if an open-architecture compression subsystem were not implemented. Closed architecture requires domestic and international standards. In order to achieve standards, international, intranational, government, and corporate agreement must be achieved. For a closed architecture, there must be international collaboration studying, implementing, and validating all current methods of algorithm coding with an attempt to achieve the best. This evaluation process is time-consuming. Later emergence of a yet better coding algorithm cannot be supported by a closed-architecture system.

Open image-compression architecture permits continued image compression algorithm research and evolution without rendering obsolete the decompression hardware sitting on a TV set or residing in a computer. Better image quality and noise enhancement pursuits may continue indefinitely with the objective of continued product enhancement by downloadable firmware. Certainly it would be arrogant to believe we have discovered the best compression algorithms at this time in history. It is very possible that image algorithm research will be taken up by movie and broadcast studios as their intellectual property (attached to their films), which permits them to deliver better images and sound than their competitors. Algorithms will be developed to permit decoding of different screen aspect ratios and resolutions, permitting HDTV, conventional digital TV, and computerized multimedia to peacefully co-exist. As we have seen, full-motion video, digital audio, and computer graphics and text are all computer data, and can be delivered over existing high-capacity computer and telecommunications networks.

More and more companies are participating in standards activities dealing with digital video. MPEG, spatial correlation, wavelet, and many other technologies all respond to that need. Which is the best technology? Which compression approach is not an imprisoning technology? Which architectural approach permits a gradual and ordered evolution of image decompression into the future? These are significant questions that must be pondered as we move into the digital interactive video revolution. Numerous image compression algorithms have been described in the literature, currently exist, and are waiting to be invented; but there is a current trend toward the discrete cosine transform in the standards industry. Certainly open architecture does not preclude MPEG; rather, it permits MPEG and its derivatives and other new compression technology to co-exist and evolve.

Some Common Compression Preprocessing Elements

Various image compression techniques use many common preprocessing facilities such as filtering, chrominance information subsampling, quantization, predictive coding, motion compensation, variable-length coding, and image interpolation, in addition to the image transformation algorithms that provide the significant compression factors. The idea is to tune all these procedures to the human vision system so no loss in image quality is perceived, while permitting the discarding of all information that is not perceptable to the human vision system.

The human vision system is most sensitive to the resolution of an image's luminance component; therefore, the "Y" luminance values should be encoded at maximum resolution. The human vision system is less sensitive to the chrominance information. Subsampling or discarding chrominance pixel values while retaining luminance values can permit information reduction with minimal degradation of perceptible image quality.

Filtering removes, attenuates, or amplifies selective information per the design of the filter. For example, a two-dimensional low-pass filter could dispose of high-frequency image content while preserving low-frequency image content, which would in turn reduce the information content. The viewed result would be a blurring of the image. Similar information reduction could be achieved by a high-pass filter that emphasizes high-frequency image content while attenuating low-frequency image content. The viewed characteristics could be a sketched-line-type drawing with little spatial gradients. More complicated filters can be fabricated to permit tailoring of the entire image frequency content, and when the complexity reaches a certain level, complex filtering is best achieved by a frequency transform function, such as one of the many transform procedures previously listed in this chapter.

Quantization is the technique for representing a sampled value with an integer code. The difference between the actual value and the quantized value is a combinition of resolution and noise. Under some circumstances, the human vision system is less sensitive to these quantization inaccuracies or noise, so these inaccuracies can be permitted without causing perceptible image degradation. This permits a further reduction in required information transmission and storage.

Predictive coding improves compression through statistical redundancy. Based on the values of the pixels previously decoded, both the decoder and encoder can estimate or predict the value of a pixel yet to be decoded, requiring only the difference between the predicted and actual values to be encoded. This difference value is the prediction error, which the decoder uses to correct the prediction. This technique is quite useful, because this also permits a further reduction in required information transmission and storage.

Motion compensation predicts the value of a cluster of neighboring pixels in an image by relocating this block of pixels to a new image using only the two-dimensional image cluster translation vectors. This technique relies on the fact that within a short sequence of images of the same general scene, many objects remain in the same location, while others may move a short distance.

Variable-length coding (Huffman coding) is a statistical technique assigning code words to a value to be encoded. Values with high frequency of occurrence are assigned short code words while infrequent occurrences are assigned longer code words, in a manner somewhat analogous to Morse code. On an average, the more frequent shorter code words dominate so much that the new code string is much shorter than the original, producing more compact transmission and storage.

Picture interpolation permits the creation of intermediate pictures. An image from the near past and an image from the near future can be combined to produce an intermediate picture, thereby (under certain circumstances) permitting reduced bit stream transmission and storage.

These techniques are useful to both transform and nontransform encoding procedures.

An Example of Transform Encoding

This section of this chapter presents an overview of the Moving Picture Experts Group (MPEG) standard. The standard is officially known as the "Generic Coding of Moving Pictures and Associated Audio" and was published June 10, 1994 as document ISO/IEC 13818-1. Since 1988, it is still in written recommendation form.

The MPEG standard addresses the compression and decompression of video and audio signals and the synchronization of audio and video signals during playback of the decompressed MPEG data. The MPEG video algorithm can compress video signals to about ½ to 1 bit per coded pixel. At a compressed data rate of 1.2 Mbits per second, a coded resolution of 352 by 240 at 30 Hz is often used, and the resulting video quality is comparable to VHS.

The activities of the MPEG committee were started in 1988 with the goal of achieving a draft standard by 1990. From 1988 to 1990, MPEG participation has increased from 15 to 150 participants and general standards were created. In 1994, the committee was arguing on details for the standard and ownership of the technology. The author acknowledges that the MPEG standard is a truly viable compression technology. It is also a valuable training tool for other image compression researchers and proponents of open decompression architecture.

MPEG has been defined as a source coding algorithm, with a large degree of flexibility, that can be used in a variety of applications permitting variations in picture sizes, aspect ratios, and bit rates. MPEG is constrained to serve pixel resolutions of less than 768 by 576, with a frame rate of less than or equal to 30 frames per second (fps).

MPEG was created to achieve a high compression ratio while preserving good picture quality. The algorithm is not lossless, as the exact pixel values are not preserved during coding. MPEG chooses techniques based on the requirement to balance high picture quality and high compression ratio with the requirement to facilitate random access to a picture database (for example, video editing purposes). Random frame access is most simply satisfied with pure intraframe coding (JPEG), but JPEG alone does not produce the level of compression often required.

MPEG has its origins in JPEG, a still-frame compression algorithm. JPEG is focused exclusively on still-image compression. JPEG could produce a series of still frames that could be viewed in rapid succession, but it would fail to take advantage of the extensive frame-to-frame redundancy present in all video sequences. JPEG takes advantage of image redundancy in two dimensions, and MPEG takes advantage of image redundancy in three dimensions. MPEG produces significantly higher compression factors than JPEG.

The MPEG video compression algorithm relies on three basic principles: transform domain-based compression in the form of discrete cosine transform (DCT) for reduction of spatial redundancy, temporal processing for reduction of frame-to-frame redundancy, and block-based motion compensation for motion prediction. Motion-compensated techniques are applied with both casual (pure predictive coding) and noncasual predictors (interpolative coding).

DCT transform

The DCT is depicted by Figure 5.1 below.

The picture elements (pixels) are shown in raster scan order or pixel space, while the DCT coefficients are arranged by frequency order. In frequency space, the top left coefficient is the DC term and is proportional to the average value of the component pixel values. The other components are called AC coefficients. AC coefficients to the right of the DC coefficient represent increasing horizontal frequencies, whereas AC coefficients below the DC coefficient represent increasing vertical frequencies. The remaining AC coefficients contain both horizontal and vertical frequency components. Note that an image containing only one horizontal line contains only vertical frequencies.

The coefficient array contains all the information of the pixel array, and the pixel array can be exactly reconstructed from the coefficient array, except for information

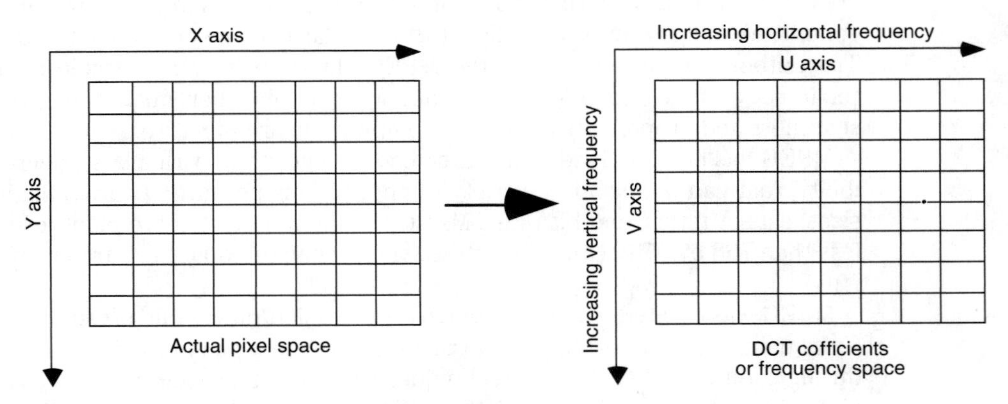

Figure 5.1 Transformation of pixels to DCT coefficients.

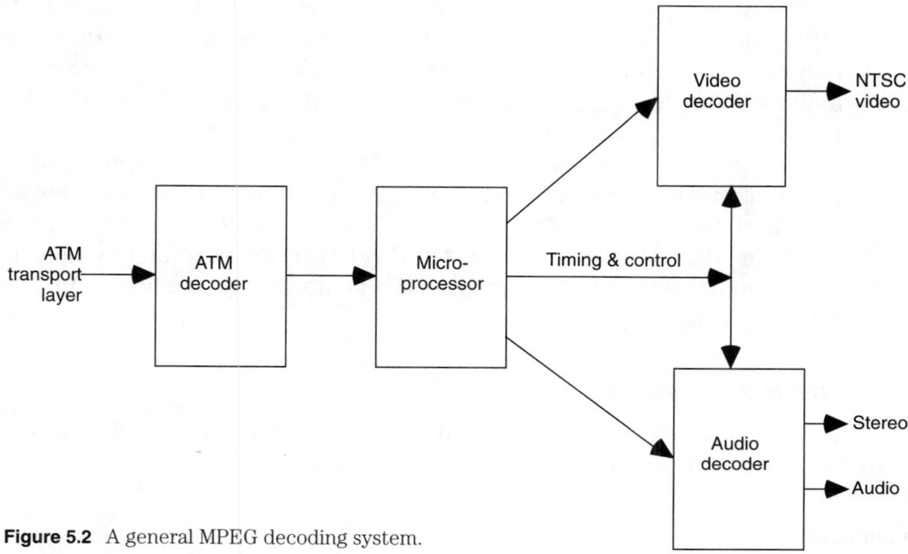

Figure 5.2 A general MPEG decoding system.

loss due to the use of limited arithmetic precision. They are one-to-one mappings of each other.

The two-dimensional DCT is defined as:

$$F(u,v) = \frac{1}{4}C(u)C(v) \sum_{x=0}^{7} \sum_{y=0}^{7} f(x,y)\cos(\pi(2x+1)u/16)\cos(\pi(2y+1)v/16)$$

with: u,v,x,y = 0,1,2,...7
where x,y = spatial coordinates in the pixel domain
and u,v = coordinates in the frequency transform domain.

It should be noted that fast DCT transforms exist and are analogous to the fast Fourier transform.

MPEG data stream structure

The MPEG stream, in its most general form, is made up of two layers:

The system layer. Contains timing and other information needed to demultiplex the audio and video data streams and to synchronize audio and video during playback.

The compression layer. Includes the compressed audio and video data streams.

Decoding process

Figure 5.2 shows a generalized decoding system for the video stream, which accepts MPEG-encoded ATM data and produces as output raw NTSC or PAL video and 2-channel stereophonic sound.

The system controller (microprocessor) extracts the timing information from the MPEG data stream and sends it to the other system components to provide audio and video synchronization. The system controller also demultiplexes the video and audio streams and sends each to the appropriate decoder. The system decoder function is implemented as software on the microprocessor.

A video decoder decompresses the video data stream as specified in Part 2 of the MPEG standard (for more information about video compression, refer to the MPEG Standard, Section 2.2, Inter-picture Coding, and Section 2.3, Inter-picture Coding). The video decoder performs the MPEG video decompression function.

The audio decoder decompresses the audio stream as specified in Part 3 of the MPEG standard.

Video data stream data hierarchy

Figure 5.3 is a schematic representation of the MPEG standard data structure hierarchy in the video stream.

Video sequence

The video data stream (sequence/flow) consists of a sequence header, one or more groups of pictures, and an end-of-sequence code. A group of pictures is defined to be

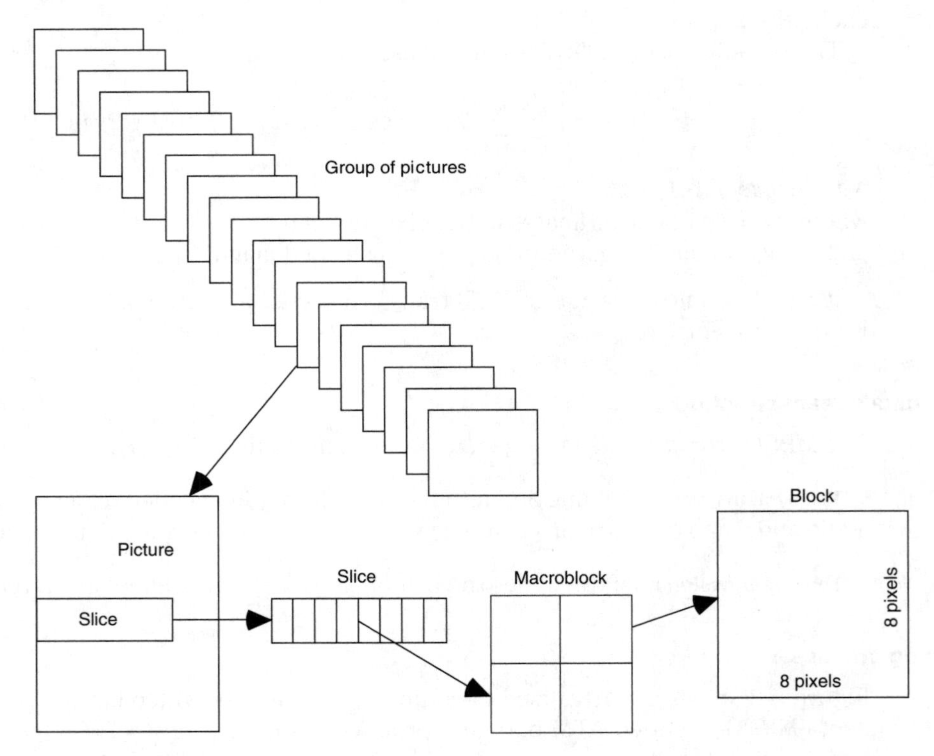

Figure 5.3 MPEG data hierarchy.

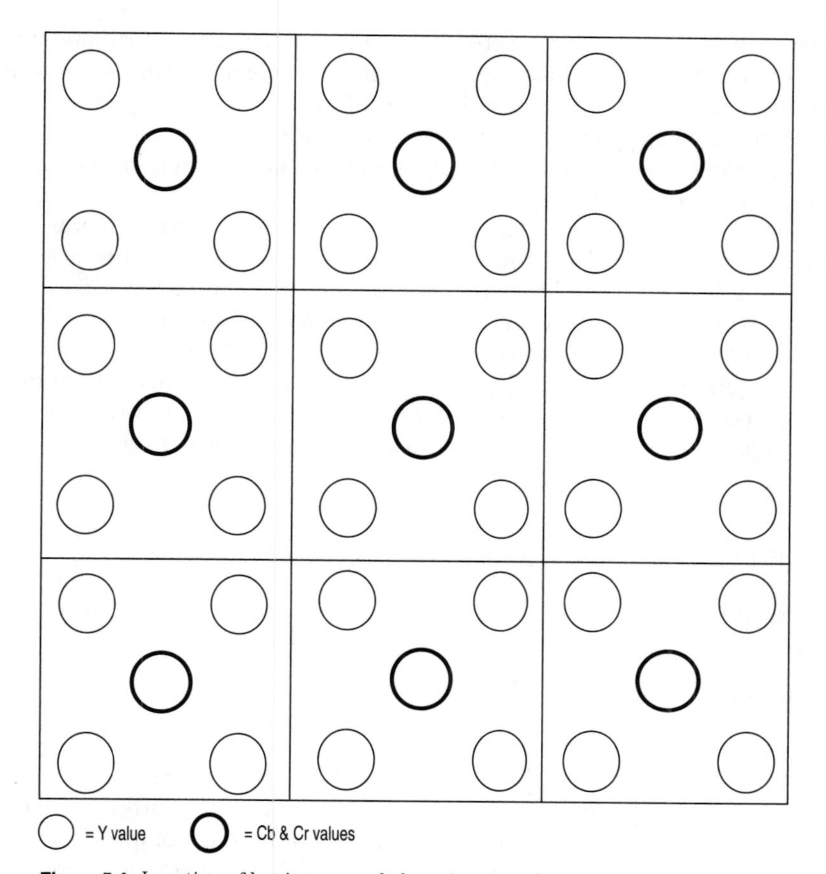

Figure 5.4 Location of luminance and chrominance values.

a series of one or more pictures intended to facilitate random access into the sequence for video editing or video-on-demand purposes.

A picture is the primary coding unit of a video sequence. A picture consists of three rectangular matrices representing luminance (Y) and two chrominance (CbCr) values. The Y matrix has an even number of rows and columns. The Cb and Cr matrices are one-half the size of the Y matrix, both in the horizontal and vertical directions.

Figure 5.4 shows the relative x-y locations of the luminance and chrominance components. Note that for every four luminance values, there are two associated chrominance values: one Cb value and one Cr value. (The location of the Cb and Cr values is the same, so only one circle is shown in Figure 5.4.)

A slice is defined as one or more interconnected macroblocks. The order of the macroblocks within a slice is from left to right and top to bottom. Slices are consequential in the handling of errors. If the bit stream incorporates an error, the decoder can jump to the start of the next slice. Having more slices in the bit stream facilitates better camouflaging of errors, but uses bits that could otherwise be used to enhance picture integrity.

A macroblock is a 16-pixel-by-16-line section of luminance components and the corresponding 8-pixel-by-8-line section of the chrominance components as represented in Figure 5.4 for the spatial location of luminance and chrominance components. A macroblock contains four Y blocks, one Cb block, and one Cr block, as shown in Figure 5.5. The numbers correspond to the ordering of the blocks in the data stream, with block 1 first.

A block is an 8 by 8 set of values of a luminance or chrominance component. A luminance block corresponds to one-fourth as large a portion of the displayed image as does a chrominance block. There is a high degree of frame-to-frame correlation, because much of the information in a picture within a video sequence is similar to information in previous and/or subsequent pictures. The MPEG standard takes advantage of this temporal redundancy to represent some pictures in terms of their differences from a reference picture. The following discussion describes the various picture types and explains the corresponding techniques used in inter-picture coding.

Picture Types

The MPEG standard specifically defines three types of pictures:

- Intra-pictures
- Predicted pictures
- Bidirectional pictures

Intra-pictures

Intra-pictures or I-pictures are coded using only information present in the picture itself. I-pictures provide random access points into the compressed video data stream. I-pictures use only transform coding, and therefore provide moderate compression. I-pictures typically use about two bits per coded pixel. The I-pictures are actual JPEG still images with no other processing.

Predicted pictures

Predicted pictures or P-pictures are coded with respect to the nearest previous I- or P-picture. This procedure is called forward prediction, and is illustrated in Figure

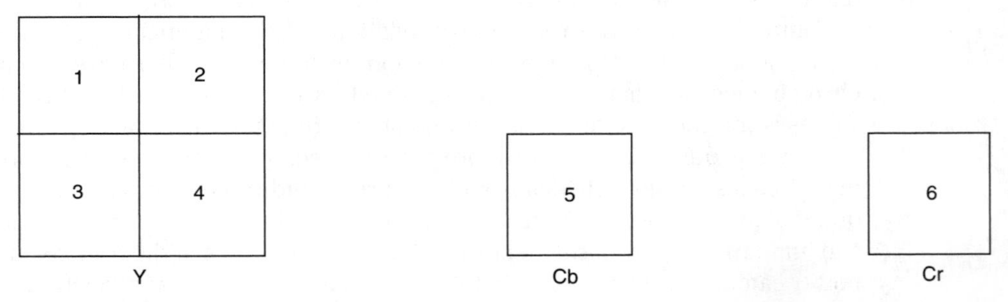

Figure 5.5 Macroblock organization.

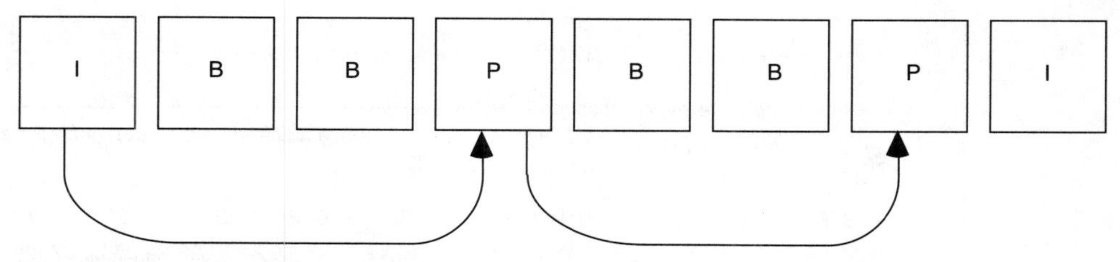

Figure 5.6 Forward prediction example.

5.6. Predicted pictures facilitate more compression and serve as a reference for B-pictures and future P-pictures. P-pictures use motion compensation to provide more compression than is possible with I-pictures. P-pictures can distribute coding errors, because P-pictures can be predicted from previous P-pictures.

Bidirectional pictures

Bidirectional pictures or B-pictures are pictures that use both a past and future picture information as a reference. This technique is called bidirectional prediction and is illustrated in Figure 5.7. Bidirectional pictures provide maximum compression and do not propagate errors, because they are never used as a reference. Bidirectional prediction also decreases the effect of noise by averaging two pictures.

Video data stream composition

The MPEG algorithm allows the encoder to choose the frequency and location of I-pictures (JPEG pictures). This choice is predicated upon the application's need for random accessibility and the location of scene cuts in the video stream. In applications where random access is important, such as video editing systems, I-pictures are typically used two or more times a second.

The encoder also selects the number of bidirectional pictures between any pair of reference (I- or P-) pictures. This choice is based on factors such as the amount of memory in the encoder and the idiosyncrasies of the material being coded. For a large class of scenes, a workable ordering is to have two bidirectional pictures separating successive reference pictures. A typical arrangement of I-, P-, and B-pictures is shown in Figure 5.8 in the order in which they are displayed.

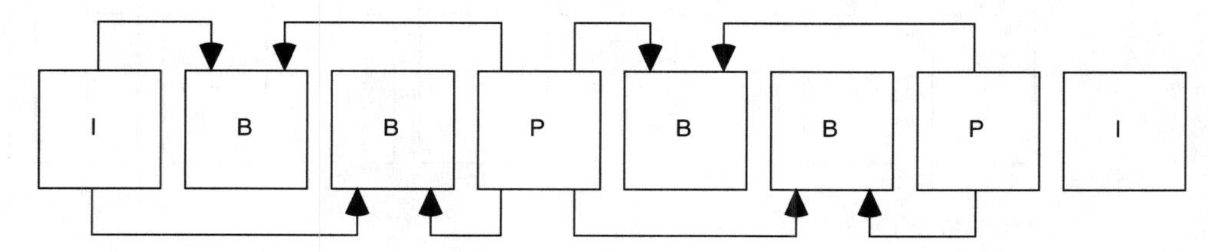

Figure 5.7 Bidirectional prediction.

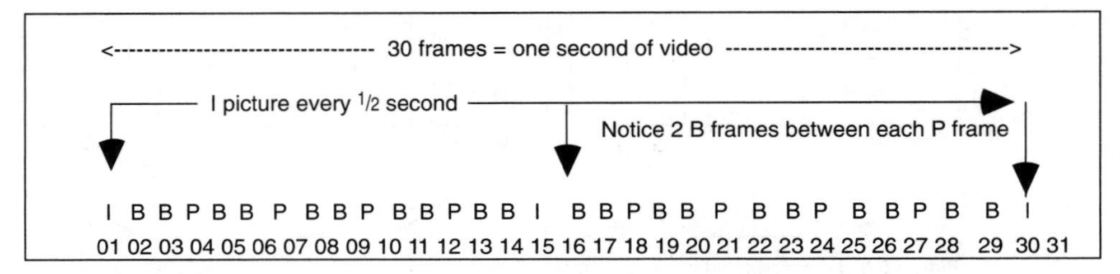

Figure 5.8 Typical display order of picture types.

The MPEG encoder juxtaposes pictures in the video stream to present the pictures to the decoder in the most efficient sequence. In particular, the reference pictures needed to reconstruct B-pictures are sent before the associated B-pictures. Figure 5.9 illustrates this sequencing for the first section of the example shown above.

Motion compensation

Motion compensation is a procedure for improving the compression of both P-pictures and B-pictures by eliminating temporal redundancy. Motion compensation

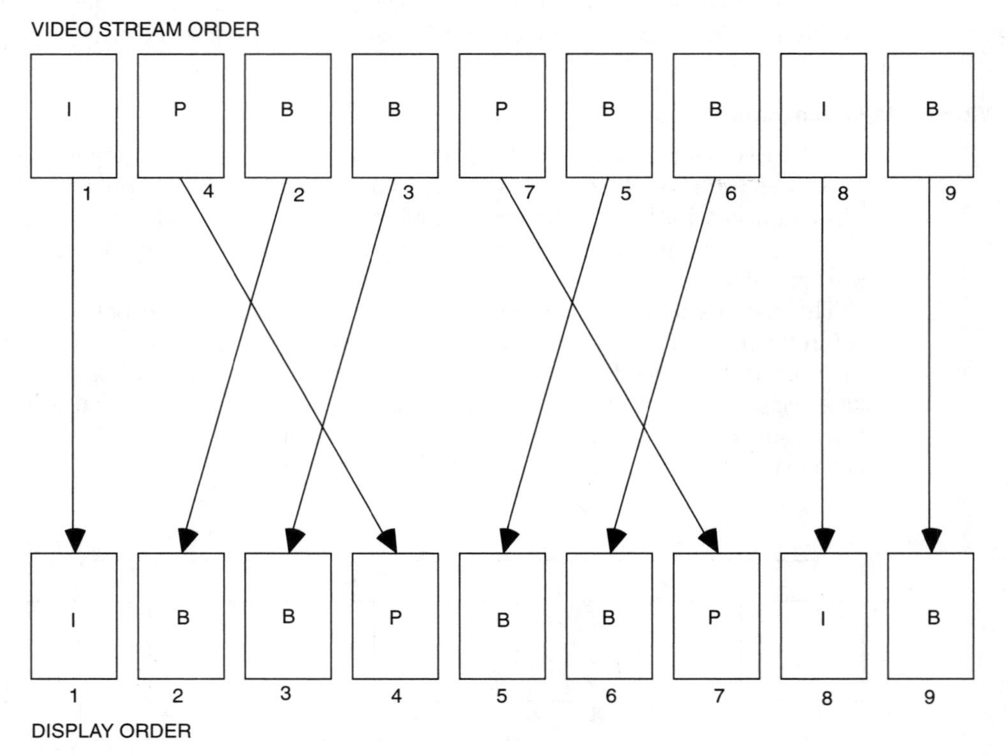

Figure 5.9 Video stream versus display ordering.

normally improves compression by about a factor of three compared to intra-picture coding. Motion compensation algorithms work at the macroblock level.

When a macroblock is compressed by motion compensation, the compressed file contains the following information:

- The spatial difference between the reference and the macroblock being coded (motion vectors)
- The content differences between the reference and the macroblock being coded (error terms)

Not all information in a picture can be predicted from a previous picture. Consider a scene in which a tree has fallen. The visual details of the scene behind the tree cannot be predicted from a previous frame in which that tree had fallen. When a macroblock in a P-picture cannot be represented by motion compensation, it is coded in the same way as a macroblock in an I-picture, that is, by JPEG transform coding techniques.

Macroblocks in a B-picture can be coded using either a previous or future reference picture as a reference, so that four codings are possible:

1. Intra-coding: no motion compensation.
2. Forward prediction: the closest previous I- or P-picture is used as a reference.
3. Backward prediction: the closest future I- or P-picture is used as a reference.
4. Bidirectional prediction: two pictures are used as reference, the closest previous I-picture or P-picture and the closest future I-picture or P-picture.

It is important to note that backward prediction can be used to predict uncovered areas that do not appear in previous pictures.

The MPEG transform coding algorithm includes the following:

- Discrete cosine transform (DCT)
- Quantization
- Run length encoding

Both image blocks and prediction-error blocks have high spatial redundancy. To reduce this redundancy, the MPEG algorithm transforms 8×8 blocks of pixels or 8×8 blocks of error terms to the frequency domain using the discrete cosine transform (DCT).

Immediately, the algorithm quantizes the frequency coefficients. Quantization is the process of approximating each frequency coefficient to one of a limited number of permitted values. The encoder chooses a quantization matrix that determines how each frequency coefficient in the 8×8 block is quantized. Human perception of quantization error is lower for high spatial frequencies than for lower spatial frequencies, so high frequencies are typically quantized more coarsely (i.e., with fewer allowed values) than low frequencies.

The combination of DCT and quantization results in many of the frequency coefficients being zero, particularly the coefficients for high spatial frequencies. To achieve maximum benefit from this, the coefficients are organized in a zigzag order to produce long runs of zeros (see Figure 5.10). The coefficients are then converted to a series of run-length amplitude pairs, each pair indicating a number of zero coefficients and the amplitude of a nonzero coefficient. These run-length amplitude pairs are then coded with a variable-length code, which uses shorter codes for commonly occurring pairs and longer codes for less common pairs.

Some blocks of pixels need to be coded more accurately than others. For example, blocks with smooth intensity gradients need accurate coding to avoid visible block boundaries. To deal with this inequality between blocks, the MPEG algorithm allows the amount of quantization to be modified for each 16×16 block of pixels. This mechanism can also be used to provide smooth adaptation to a particular bit rate.

Timing and Control

The MPEG standard provides a timing mechanism that ensures synchronization of audio and video. The standard contains two parameters, the system clock reference (SCR) and the presentation time stamp (PTS).

The MPEG system clock running at 90 kHz generates 7.776×10^9 clock cycles in a 24-hour day. System clock references and presentation time stamps are 33-bit values, which can represent any clock cycle in a 24-hour period.

System clock references

A system clock reference is a snapshot of the encoder system clock. The SCRs used by the audio and video decoder must have approximately the same value. To keep their values in agreement, SCRs are inserted into the MPEG stream at least as often as every 0.7 second by the MPEG encoder, and are extracted by the system decoder and sent to the audio and video decoders as illustrated in Figure 5.11. The video and audio decoders update their internal clocks using the SCR value sent by the system decoder.

The MPEG standard provides a timing mechanism that ensures synchronization of audio and video. The standard includes two parameters used by the decoder: the system clock reference (SCR) and the presentation time stamp (PTS).

Presentation time stamps

Presentation time stamps are samples of the encoder system clock that are associated with some video or audio presentation units. A presentation unit is a decoded

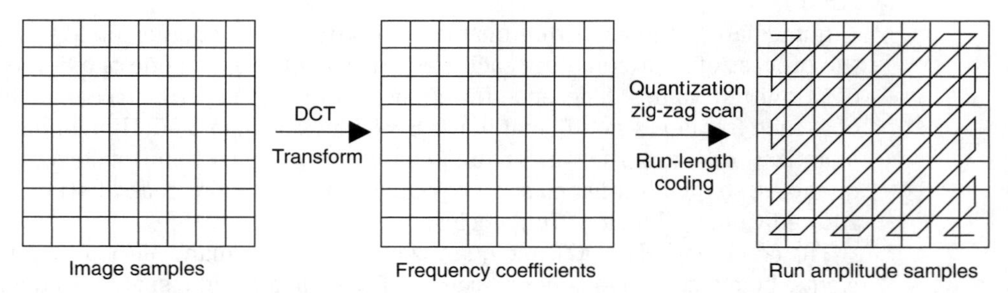

Figure 5.10 Transform coding operations.

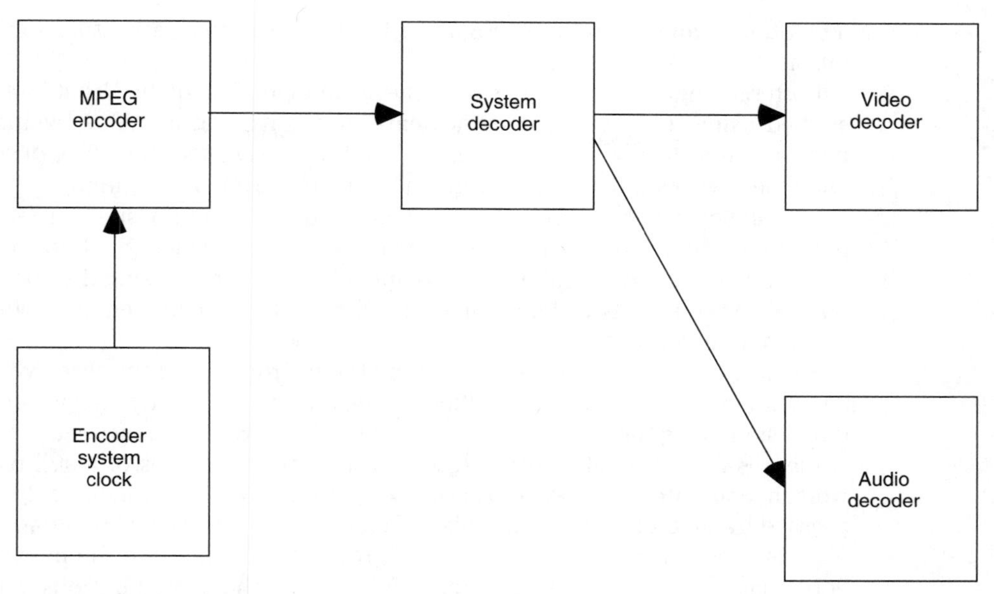

Figure 5.11 SCR flow in MPEG system.

video picture or a decoded audio time sequence. The encoder inserts multiple PTSs into the MPEG stream at least as often as every 0.7 second. The PTS represents the time at which the video picture is to be displayed or the starting playback time for the audio time sequence.

The hardware decoder either deletes or repeats pictures to ensure that the PTS matches the current value of the SCR when a picture with a PTS is displayed. If the PTS is earlier (has a smaller value) than the current SCR, the decoder discards the picture. If the PTS is later (has a larger value) than the current SCR, the decoder repeats the display of the picture.

An Example of Nontransform Encoding: Polyhedral Encoding

The primary representation of spatial objects may be defined by small geometric figures consisting of clustered pixels whose correlation coefficients exceed a certain threshold value. In order for these geometric forms selected to be easy to manipulate, their complete description must be concise and controllable. Polyhedral encoding includes temporal space as the polygon's third dimension. Therefore, the polyhedral representation includes interframe encoding implicit in it.

As is the case for other image compression procedures, polyhedral representation of image elements requires significant computation to produce compact data structures. Certain geometric forms, however, have great advantages in the ability to represent large volumes in a very compact format. In particular, representation by rectangular solids and triangular prisms are easily represented and easy to manipulate by a receiving subsystem. Further, because any polyhedron may be described as a collection of triangles, there is no limit to the complexity of an object when triangles are used. Figure 5.12 shows the basic structure for representation.

Both rectangles and triangles have the advantage of "good fit." That is, a space described using these forms does not have any remaining undefined volumes to be filled by interpolation or other methods by the receiving system. The description is complete. Both allow continual refinement to any level of resolution.

Rectangles may be defined by the coordinates of two opposing corners. This requires two values to define an area of any size from a single pixel up to the entire screen. A rectangular solid may be completely specified by using the coordinates of one corner and the edge length in each of three dimensions, requiring three values for complete description.

Similarly, triangles may be specified by the coordinates of the three vertices, and a rectangular prism may be specified by the addition of an edge length in the third-dimension, or as the upper or lower triangle of a represented rectangle.

Figures 5.13 through 5.16 represent various process steps of a still picture of a woman, computer encoded as a series consisting only of rectangular polygons, compressed by a factor of 0.6 bit per pixel. Figure 5.13 is the original photograph. Figure 5.14 uses pseudocolor (represented by greyscale) to illustrate the individual polygons that make up the image. Figure 5.15 shows the same polygons with natural color (grayscale), and Figure 5.16 is Figure 5.15 enhanced via filtering. No subsequent image enhancement has been applied.

Combination of shapes

Polyhedral encoding can mix rectangles and right-angle triangles (and the corresponding solids). Because any right-angle triangle may be described as either the upper or lower portion of a rectangle, this has the advantage of the rectangular format with the added flexibility of triangles for irregular objects. Only two additional bits are required for specification of the object type. The format could be as follows:

type code ¦ *vertex* ¦ *edge length* ¦ *data*

where

type code is a two-bit indicator as follows:

00 = rectangle
01 = upper triangle
10 = lower triangle
11 = reserved

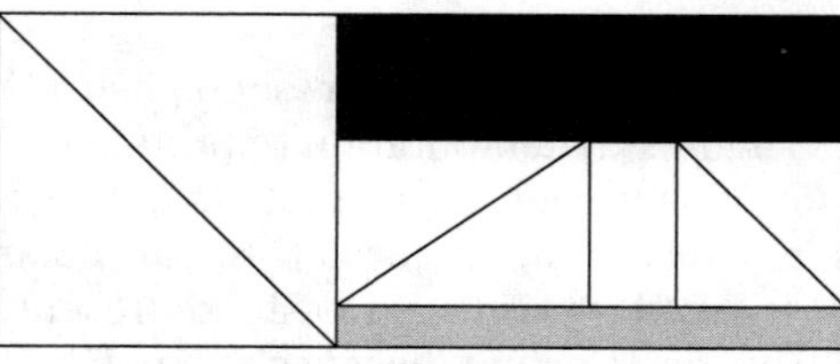

Portion of Image Measuring
25 pixels by 10 pixels

Figure 5.12 Illustration showing how triangles and rectangles can be grouped together to cluster highly correlated pixels.

Figure 5.13 Woman polyhedrally encoded to 1.2 bits per pixel.

Figure 5.14 Woman polyhedrally encoded to 0.9 bits per pixel, represented by pseudocolor.

Figure 5.15 Woman polyhedrally encoded to 0.6 bits per pixel.

Figure 5.16 Woman polyhedrally encoded with digital enhancement.

vertex is the specification of the upper left corner of the rectangle in three dimensions.

edge length is the length of an edge in the third dimension.

data is the image data, which may be a color/luminance value or a pattern specification.

The process of implementing the polyhedral encoding approach requires specification of proper "objects" for representation. The simplest method to implement is to create an "object" for each area of a discrete color/luminance value (or its nearest neighbor whose correlation coefficient exceeds the required threshold) to be an object and to subdivide this object until it is completely specified by one of the three primitive forms (upper/lower triangle or rectangle).

When image complexity exceeds the channel bandwidth facilities, the inter-pixel correlation coefficient thresholds can be adaptively reduced until the bandwidth requirements meet the channel bandwidth. The visual impact of this approach produces slight image degradation and, in extreme cases, a tiled-image effect appears. Subsequent image enhancement can remove most of these visible artifacts.

Movement of these three basic object types is also easy to represent. All that is required is an offset along one or more of the three axes. The receiving subsystem then performs a simple addition to produce a new specification for the corner of the resulting rectangle.

The objects may also be grouped in a hierarchical manner. That is, a change may apply to a section of the image that is composed of several objects. An encompassing rectangle may be specified along with an operation to be performed (overall change in luminance, displacement, etc.) on all objects or partial objects within the area indicated.

A curious aspect of the approach is the specification of movement that results in the overlap of two or more objects. The intersection or difference of two objects can produce multiple objects of the three primary types. This is prevented by a change operator that is employed in both the receiving subsystem and the reconstruction process in the transmitting subsystem that produces the actual object definitions.

In summary, the approach to encoding using the three primitive object types described has relative appealing simplicity and performs reasonably well.

Open Image-Compression Architecture

It has been shown that a number of good image compression algorithms exist, and this chapter has only touched on a few of them. Undoubtedly, new and better algorithms will continue to be invented, some having better temporal properties, some having better spatial processing properties; some may even produce more vivid colors with higher resolutions, some may operate better with higher resolutions, some may work better with scenes that have high motion content, some may be more impervious to noise, etc. It may be desirable to encode different scenes with different compression algorithms. Of course, dynamic decoding of these different compression algorithms would require dynamically reconfigurable and open image-compression architecture.

Perhaps such a dynamically reconfigurable environment is not required, but the requirement for an open image-compression architecture exists. If set top box obsolescence is to be avoided, open architecture is required. Before any image compression hardware is designed, one should ask the question "what is common to all image processing algorithms?" to determine what hardware can be optimized for the general solution.

In this chapter, we have seen that in all compression algorithms reviewed that a continual examination of pixel groups is required. Experimentation has shown that hardware specifically tuned to examine 2 or 3 dimensions of pixel clusters along with good computational capability can become a formidable general-purpose digital image processor (DIP). The subsequent figures illustrate how such an open architecture device could be implemented in VLSI hardware.

A firmware microprogrammable emulator-style architecture, such as is employed by the x86 and 68xxx microprocessors, provides a good starting point for DIP architecture. Such a single data flow (emulator slice) is shown as Figure 5.17. Its differences from conventional microprocessor architecture are immediately apparent by the dedicated microflow registers used to achieve tightly coupled multiprocessing with other slices as shown in Figure 5.17, and its unique microprogram branch control circuits shown in Figure 5.19.

The portion of the architecture represented in Figure 5.17 requires less than 80,000 gates for implementation; therefore multiples of these slices or tightly coupled microprocessing elements can be fabricated onto one VLSI device to provide very high performance. Because of the nature of images, different processors on one VLSI chip can compress different pieces of the image, and using stochastic pixel cluster selection procedures by dividing an image into "N" pieces, these pieces when

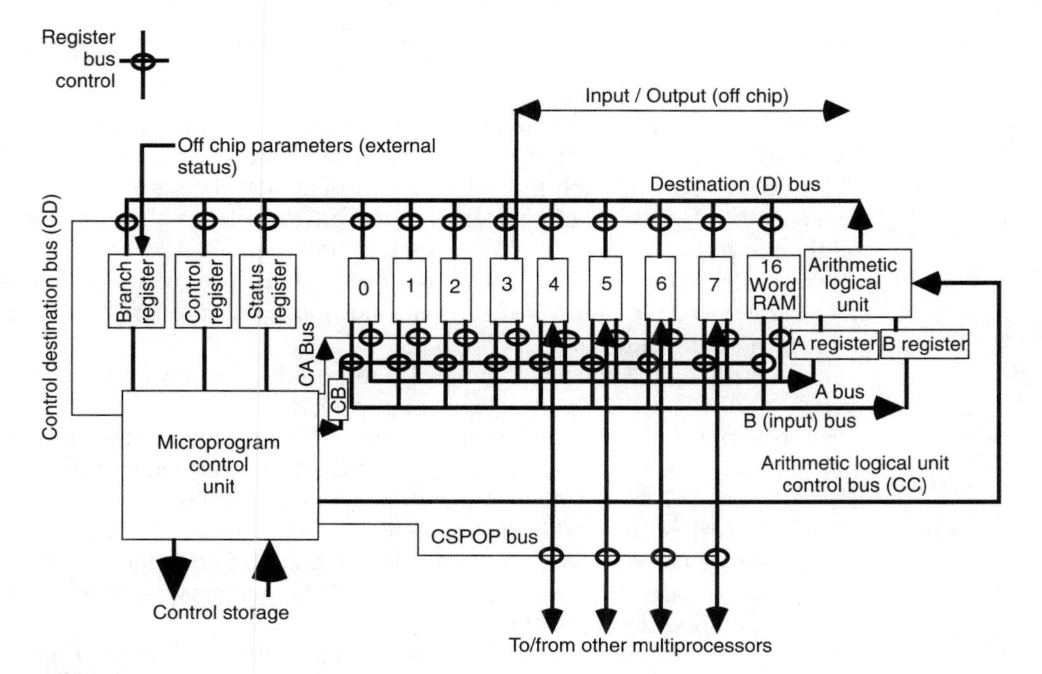

Figure 5.17 VLSI replicable microprocessor slice.

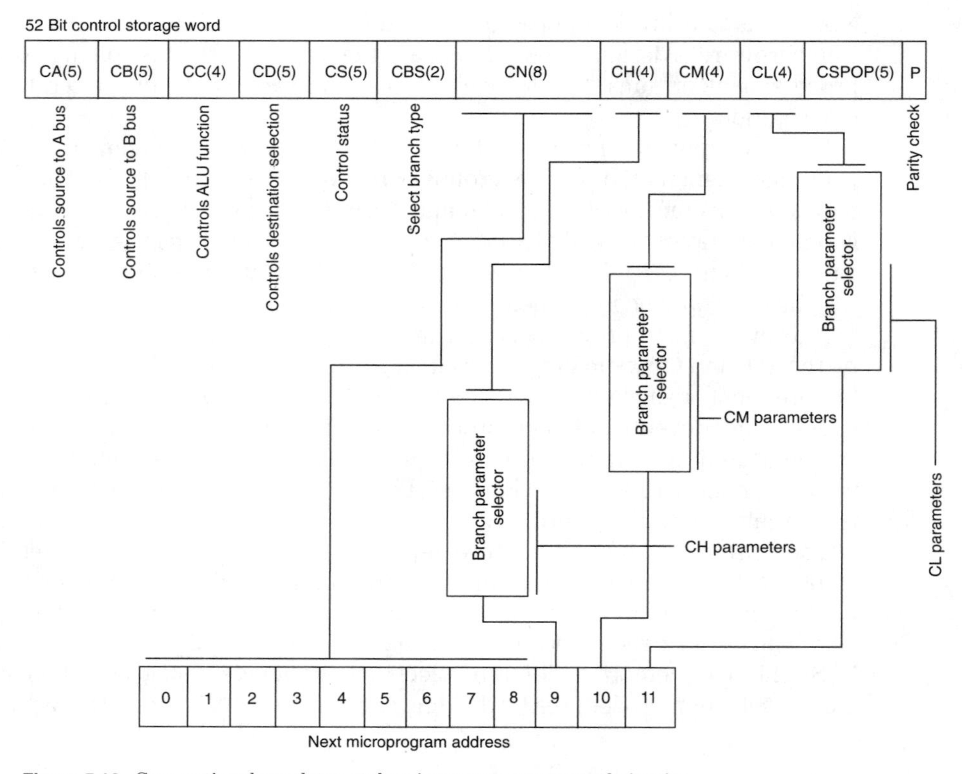

Figure 5.18 Conventional emulator-style microprocessor control circuits.

combined will not produce visible artifacts because they will have different spatial characteristics and geometries on a frame-to-frame basis.

The unique part of each individual microprocessor slice is its control circuits. Each slice has two types of control circuits, conventional as shown in Figure 5.18, and the image-tuned control circuits as shown in Figure 5.19.

Figure 5.18 represents the control paths of each microprogrammable element of the VLSI device, the control of the data flow, and microprogram instruction sequencing and branching.

This control word, shown in Figure 5.19, represents the flow for tuned pixel processing for JPEG, MPEG, Spatial Correlation, MIT, 3DRLL, etc.

Multiple, tightly coupled microprocessors will be designed onto a single large-scale gate array, as shown in Figure 5.20, but the heavy dashed and the heavy solid paths represent data and control for attachment of shared external memory and shared onboard content-addressable memory, respectively. The shaded paths are shared register paths. This architecture features automated memory collision prevention and a shared 256-word content-addressable memory suitable for accelerating image compression algorithm execution.

A physical representation of a multiple tightly coupled multiprocessor VLSI arrangement is shown in Figure 5.21.

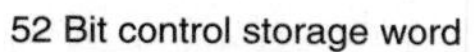

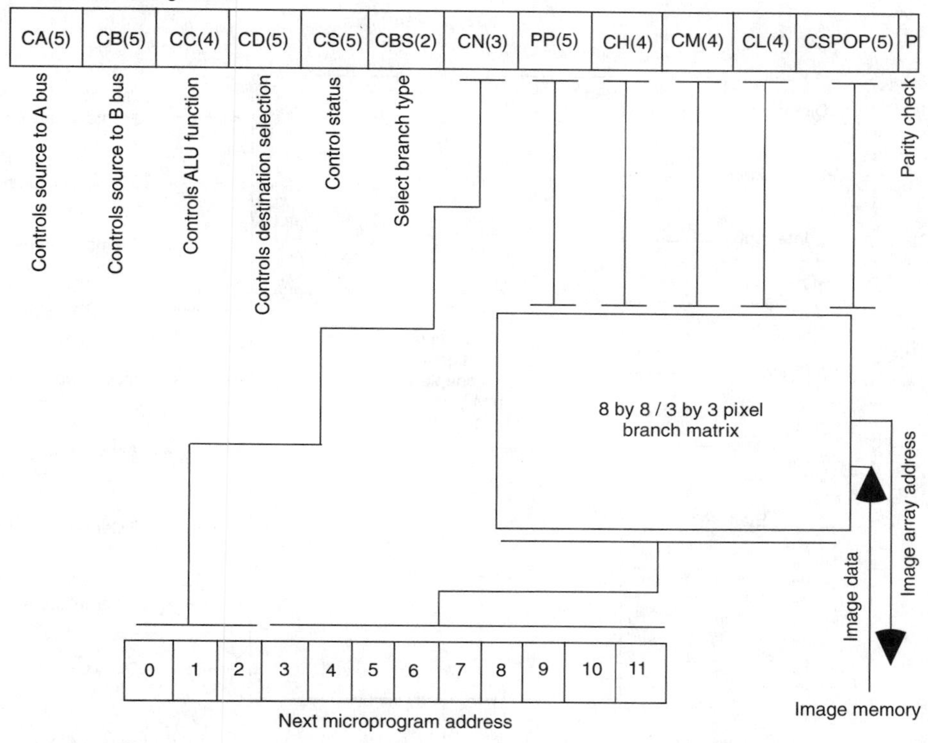

Figure 5.19 Microprocessor microprogram control circuits tuned for image processing.

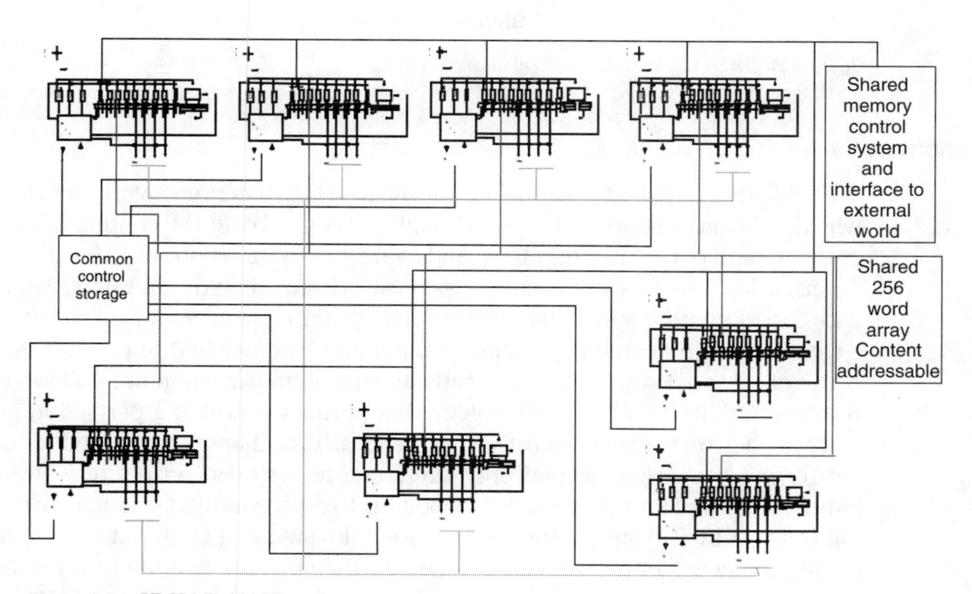

Figure 5.20 VLSI MP SLICE paths.

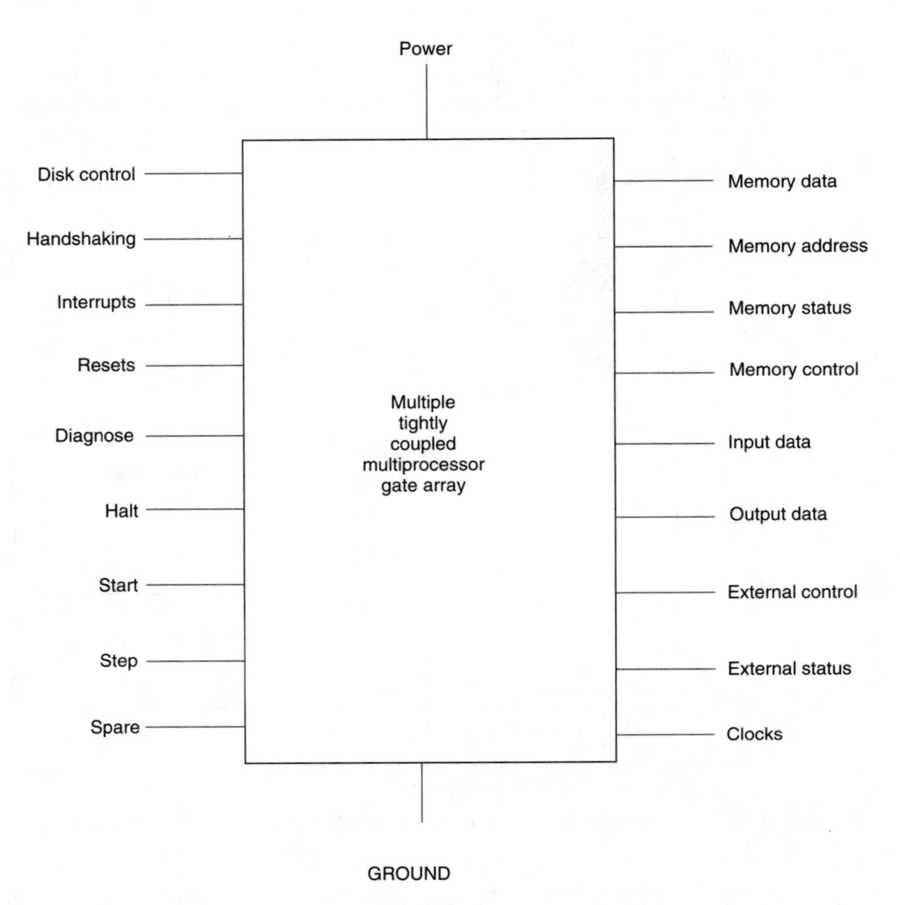

Figure 5.21 The VLSI MP DIP physical chip packaging.

Conclusions

This chapter has asked questions about the sanctity of certain algorithms. It has asked why the media creators and artists should lose control over the content of their artistic creations and permit perturbations and blemishes to this content. It questions whether "a final selection of a specific compression algorithm" should be permitted. It states the case for continuing research and development into image-compression technology. It suggests that an open compression architecture be employed to prevent the obsolescence of set top boxes, TV sets, and advanced multimedia computers. This chapter has revealed a philosophy and technology that permits continuing perfection of imaging technologies that minimize cost consequences. This chapter has intended to make us aware that properly designed open-architecture systems can permit continuing enhancements well into the future. The point is to design a smart technology without being trapped by it. Plan for the future by employing an open-architecture philosophy. Satisfy the needs of the media producers so they can provide the best product to the end user.

Chapter

6

Interactive Television and Consequential System Requirements

The more interactivity in an ITV system, the busier the resources, which include the set top box, the communications channel, and the head end equipment, usually the video server. When the set top boxes are not being used by the viewers, the system is relatively peaceful. The set top box is always powered on, listening to a designated control channel and responding to status requests and control functions. When the user is not viewing or participating in an ITV scenario, the system, while operational, is quite peaceful. Conversely, when all the users are actively viewing ITV-mode programs, the system is quite active.

The head end can address specific set top boxes, provide control, and collect status continuously. This provides an insignificant amount of traffic on the communications channel, but it provides a means for continuous system testing of all communications facilities, all set top boxes, all repeaters, all nodes, all hubs, and all circuits to and including the head end. The video server "test" dialog is typically:

- address the set top box
- request the set top box status
- send test message to the set top box
- request return of the message
- validate message integrity
- go to next set top box address
- repeat process

The system integrity using ITV technology increases from the previous CATV approach, where a dissatisfied customer calls in a maintenance report, to the fully automated system where the system operator knows about failures before his customers do, giving him extra time to correct the problem. A broadcast scenario exists where the head end transmits menuing and navigation information and special information to all subscribers' set top boxes simultaneously.

The set top box becomes active (to the subscriber, but not to the network) when the subscriber decides to interface with the box. The box has a small amount of memory to store and present messages and menus, and to control navigation through those menus. When the subscriber has navigated through the initial program selections, the set top box creates a message to be transmitted to the head end using out-of-band (oob) signaling (likely in the 5–45-MHz frequency range) that describes the customer's request. The request could be for a movie, fast food, home shopping, etc. The system returns an acknowledgment message to the subscriber and if the request was for a movie, arrangements with the set top box are made to send the movie. The oob signaling can be maintained efficiently as very small packets identifying the subscriber (by his or her box serial number) and his or her requests require very few bits and resulting small bandwidth allocation. A single oob signaling path may be quite applicable for systems with even large numbers of subscribers, because the communications occurrences are infrequent. It is possible on a very large system with many set top boxes that a backlog queue may occur for a couple of reasons: insufficient channel capacity, and/or insufficient billing computer capacity.

Either dilemma can have the same solution, namely to permit the set top boxes to freely decode the program material for a few minutes until the system can respond, then permit viewing to continue or not. This short time interval will give the system added time to redistribute resources, permitting short duration peak loads to be averaged over longer periods of time. There is no negative consequence to the subscriber, unless he or she is not authorized to watch the movie (probably for nonpayment of bills) and he or she sees a few minutes of the movie only to have it turned off. This could be thought of as a free few minutes of program preview.

By appropriate system design, perceived bottlenecks can be circumvented. By allowing set top box status to be stacked for subsequent and later processing, higher system responsiveness will occur. The appropriate system design not only affects the server and set top boxes, but it affects the limited bandwidth cable or fiber facility. Efficient bandwidth utilization is necessary if a finite number of programs and finite number of retransmission intervals are going to exist. A single coax might transmit 700 program elements or threads, but that may be only 35 movies with 20 different start times.

The system can gain efficiency if each program thread is represented by one virtual address that the set top box decodes. In this way, many set top boxes can be simultaneously decoding one thread of one movie (using the same virtual address) in only a fraction of the bandwidth if separate data packets of the same thread data were sent to multiple set top boxes. The system can uniquely address each set top box with its real address (electronic serial number) and give it an interim shared virtual address for use during movie decoding. Because these virtual addresses are known *a priori*, temporary authorization to use these addresses can be provided by the system during times of overload. When overload peaks cease to exist, permanent authorization to use these virtual addresses will be given.

The system head end can provide a series of commands to any or all of the set top boxes, and the set top boxes can make requests to the head end. Examples of head end commands include:

- set time (each set top box maintains its own time until set)
- set virtual address (each set top box decodes real addresses until virtual address is received)
- collect billing information (each set top box retains current billing information)
- request status (set top boxes continuously collect operational status)
- diagnose set top box (set top box collects malfunction information)
- diagnose system (set top box and head end test communications channel)
- set restrictions (permit only certain types of programs to be viewed)
- control customer TV (turn on/off, select channel, volume, surround sound, etc.)
- control customer VCR (turn on/off VCR, command record/play/rewind)
- initialize set top box (send all new parameters to set top box)
- reset set top box (cause set top box to go through power on restart procedure)
- disable set top box (to accommodate nonpayment problems)

These commands can be broadcast or selectively addressed. While they may not appear as shown in Figure 6.1 below, they probably will be shipped as standard 53-byte ATM packets with a 5-byte header and 48-byte data packet. Figure 6.1 is shown merely to illustrate one possibility. What is certain is that when programs are being transmitted, the set top box will be busy examining virtual addresses in the header, extracting video and audio from the rest of the packet, and processing it. The bulk of the channel bandwidth will be used for video and audio transport while control, diagnostics, initialization, and special functions will represent infrequently transmitted packets.

The head end to set top box communications is a fixed 53-byte packetized communications protocol, and many channels for transmitting these packets will exist; see chapter 10 (Communications).

When the set top box requires communications with the head end, it creates an attention interrupt. Only *one* initiating response from the set top box to the head end is

ATM Packet From Head End to Set Top Box

Header	Type	Audio and video
		Control
		Diagnostics
		Initialization
		Special function

Figure 6.1 Hypothetical ATM transmission packets from head end to set top box.

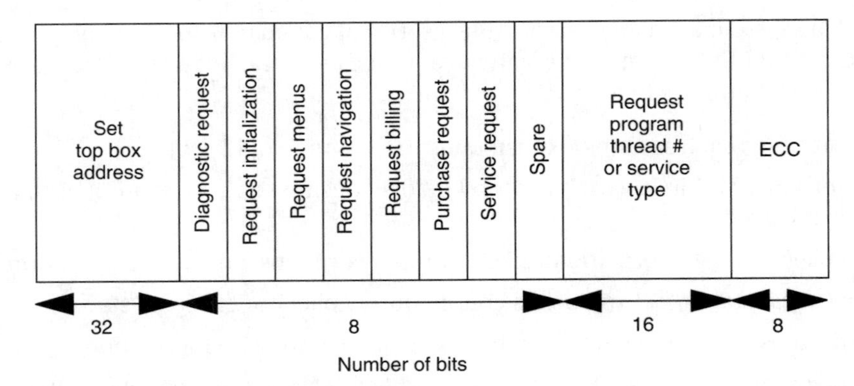

Figure 6.2 Hypothetical set top box oob packet.

necessary to create a variety of scenarios, including a request for a video. The smaller the number of unsolicited requests, the smaller the channel bandwidth required. When the set top box requires connection, it sends an attention interrupt to the head end requesting communications. That attention interrupt hypothetically could appear as Figure 6.2, and could include any of the following requests, singularly or plurally.

- diagnostic request
- request initialization
- request menus by layer
- request navigation information
- request program thread
- request billing information/account status
- request service
- request alternative telephone dial tone

A 64-bit set top box-to-head end packet seems adequate to achieve the system requirements. The purpose of examining packet possibilities was to acquire a rough idea as to what the ultimate number of bits per second down the oob (set top box-to-head end) channel might be and how long it might take the system to process this much traffic.

Let us assume a system with 10,000 subscribers (10,000 set top boxes) and one oob channel with 40 MHz of shared bandwidth. Further, assume QAM encoding at 4 bits per hertz, but only a 10% traffic capacity due to collisions, routers, switches, computer latencies, etc.:

40-MHz QAM bandwidth implies 160 Mbps.
10% utilization implies 16 Mbps.

A 64-bit packet implies 16M/64 packets per second = 250,000 oob packets per second.

If all 10,000 subscribers requested programs at exactly the same time (a highly unlikely event), then most requests would be required to be queued by the set top box until the head end was able to process them. This would require:

10,000 /250,000 or 40 ms to communicate all requests.

The following table illustrates set top box communication wait time when all subscribers make instantaneous requests, for systems with different numbers of simultaneously requesting subscribers.

While these numbers are based on worst-case assumptions, they indicate that this procedure is viable. These numbers further indicate that part of this oob spectrum could be shared as a broadcast control channel, similar to cellular phone system control channels. This channel could be used for directing set top boxes to present subscribers alternative video material while their time in the waiting queue is being processed.

We have looked only at channel induced delays so far. What is the effect of billing computer and video server delays? At the head end, equipment exists to provide the billing function, control function, and video selection/transmission function. Assume a functional electronic building block exists that extracts the oob packets from the set top boxes at the head end, and that it can route all packets to a billing computer where authorization and billing take place. Instantly, before validation occurs, a message can be transmitted to the set top box with temporary decoding information that will be valid for a few minutes. This will give the billing computer a few minutes of time to process large waiting queues while providing the customer a seemingly immediate response.

It is common to hear of a large mainframe processing hundreds of database transactions per second. At this rate, 10,000 transactions would require tens of seconds. On a Pentium disk-based transaction processing system, 5 transactions per second would translate into 2000 seconds, or 33 minutes. The secret of using a Pentium-based processor is to maintain the most important part of the database in RAM, minimizing current disk activity and deferring transaction updating to non-

Number of subscribers	Channel wait time in seconds
1,000	0.004
2,000	0.008
4,000	0.016
6,000	0.024
8,000	0.032
10,000	0.04
20,000	0.08
50,000	0.2
100,000	0.4
1,000,000	4

Figure 6.3 Table represents channel wait time as a function of number of subscribers simultaneously making requests from their set top boxes.

busy times (e.g., late at night) or during other idle times. Assuming a customer's active RAM-resident database segment is 128 bytes, a 10,000-customer RAM-resident database would require only 1.28MB of RAM for the customer database. The film database for frequently requested films, including thread timing information and assuming 128 bytes per film, would require only thousands of bytes of RAM. Processing 100 transactions per second, 10,000 transactions would require only 100 seconds to completely process in RAM. By staging the processing into two portions, preliminary authorization and final authorization, it is possible to reduce initial authorization to the set top box to a few seconds using low-cost Pentium-based microcomputer systems.

The preliminary authorization would require no user validation, merely the temporary assignment of a program virtual address to a specific set top box. The set top box would reset this temporary authorization if permanent film authorization was not received within 3 minutes. Because nonpaying customers boxes would be programmed off during nonbusy times until payment was received, the risk of a nonpaying customer seeing a few minutes of a pay program is minuscule.

The secondary authorization occurs after the customer request has been updated and approved from the RAM-resident database. This may occur immediately, or up to a few minutes later in the case of a very large and busy system. During billing computer idle times, the RAM-resident database is used to update the disk-based database.

We have discussed a simple single-hub CATV or TELCO video delivery system. There are benefits to providing resources belonging to one hub to viewers of another hub, because the different resources become additive and customer choices are increased. At the same time, additional switching facilities and more communications bandwidth are required. Accounting becomes more complex. If system A is providing a film to a viewer on system B, where is the database maintained and how is it maintained? Does system A need to have a larger RAM database to maintain the same level of responsiveness, and when and how are the databases updated? The cellular telephone industry has the same problem for "roaming" subscribers, i.e., subscribers from one system using the resources of another. There are multiple solutions to these types of problems, and this is where standardization needs to occur.

Subsequent chapters of this book examine multiple-hub system architecture, but this chapter only acknowledges the operational and billing problems. When all the CATV and TELCO systems become interconnected to become the information superhighway, these rules and regulations must be resolved. It is likely, following the cellular telephone example, that these procedures will evolve as systems begin the process of interconnection. Different interconnected groups may use different procedures, but as they combine into larger groups, standardization will be required, starting with small geographical areas, building into larger geographical areas, and finally impacting international areas.

Chapter

7

True Video on Demand vs. Near Video on Demand

Note: This was a chapter delivered to the National Cable Television Conference May 24, 1994 by Winston Hodge, Hodge Computer Research, and Chuck Milligan, Storage Technology Corporation.

Video on demand (VOD) has the potential of giving individual television viewers nearly instant access to a wide range of recorded movies, video programs, games, information, and other services. It is distinguished from more conventional TV viewing by a high degree of interactivity between the viewer and the material being viewed.

A perception exists in this industry today that each person interacting with his or her TV demands instantaneous response. This is called true video on demand (TVOD). As this chapter will show, TVOD is extremely expensive when it provides for all services possible.

The alternative to TVOD is near video on demand (NVOD). This chapter will demonstrate that while NVOD is significantly less expensive to implement, an NVOD system can be designed so that its delays are not objectionable to the user for many applications. Procedures and strategies for concealing customer latency time will be described, along with the cost differential attendant to eliminating it.

Access to recorded material with zero access time is not physically possible. Fractional second access is possible, but would be very expensive for an unlimited menu of choices by an unlimited number of subscribers.

Clearly, the quantification of cost to provide service versus the latency time is of serious importance. But there is more to the implementation decision than cost. The psychological effects of waiting come into play. For example, is one second too long to wait? How about two seconds? How about two minutes? All things being equal (which they are not), the shorter the service time the better.

This chapter will provide a clear view of physically possible service times and the cost to provide those services using advanced technology hierarchical storage. A

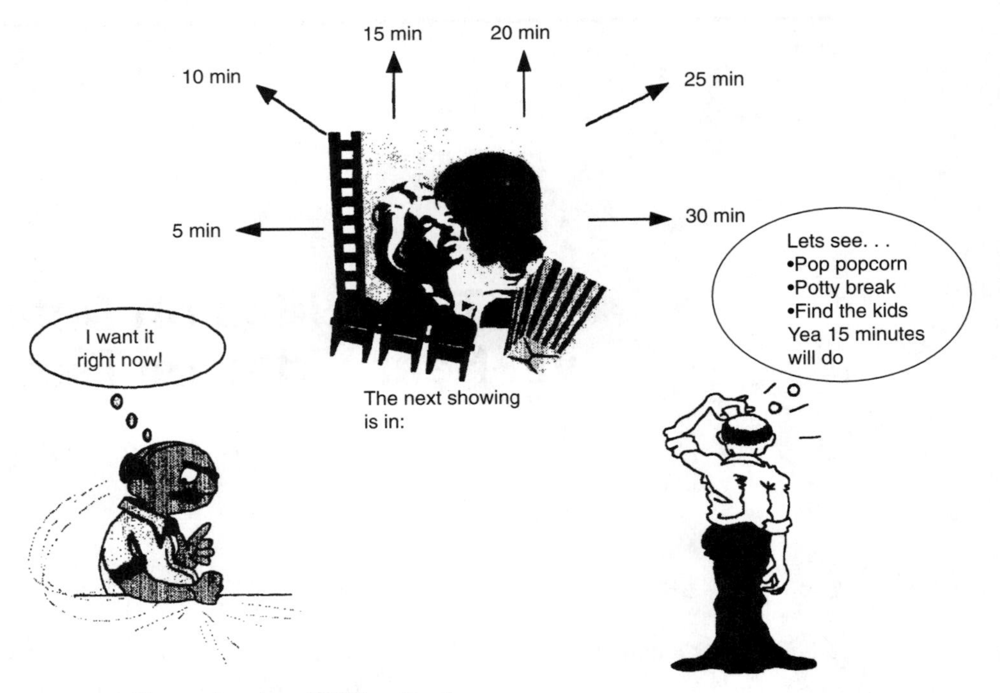

Figure 7.1 Time options in an NVOD application.

model will be described that demonstrates how the system cost varies with viewer latency. This model will be applied separately and collectively to the video servers, disk storage complexes, large terabyte robotic tape farms, VOD selector switches, communications channels, and viewer selection mechanisms.

Block diagrams used in the systems analysis and simulation will be included, along with charts and graphs which will clarify the results of the analysis. The chapter will conclude with recommendations for an economically viable system design.

Definition and Requirements

Video on demand (VOD) trial systems in one form or another are currently being implemented. An understanding of the cost factors related to response time (i.e., viewer selection latency) will provide insight into the overall system costs.

Interactivity is much more than channel selection. It may be the simple ability of the viewer to decide *what* program he or she wants to watch, and *when* he or she wants to watch it. It might allow him or her to select from among several different endings to a movie thriller. It may allow him or her to take a simulated walk down a supermarket aisle *he or she* selects, ordering products from among those displayed. It could allow him or her to engage in a simulated trip through the solar system or through a Mayan temple, making decisions about which planets to explore or which corridor to turn down, through the wonder of virtual reality. It could even allow him or her to engage in a simulated dogfight with another viewer through an interactive video game.

The foregoing scenarios require progressively increasing levels of interactivity. The response times required of the system also vary widely between the applications. For example, when home shopping, the response time from advertisement to order placement is not critical, but the navigation response from product to product is more significant. The viewer may be more concerned about the time between selection and delivery of a new movie, but whether this time interval is fractions of seconds, seconds, or even minutes may not be consequential.

A video arcade game or a virtual reality session requires much more rapid response—far beyond the capabilities of even a very large mainframe computer to service a large number of clients. For these applications, the interactivity will be supplied by downloading a program to a set top box for execution. Given this fact, once more the initial response between ordering the game and its actual delivery falls into the same degree of urgency as the ordering of a movie.

Selection time is subject to the laws of physics. These laws place limits on what it is physically possible to achieve. By knowing where the limits are, and by understanding the cost of approaching these limits, one is in a position to make objective decisions on implementation approaches. This chapter will enlighten the reader with the options currently available.

Strictly speaking, true video on demand requires instantaneous response, probably less than a second from the time a program request is made until the time the program is delivered. This has significant cost ramifications not only for the video server and video disk drives, but for the communications channel and other system elements not addressed in this chapter.

Near video on demand requires only a reasonable and convenient response time from program selection to program delivery. This interval could range from seconds to a few minutes or in some cases even a few hours. During the interval, stock material (such as seen in theaters) or interactive advertising for food or other products to be delivered to homes or music video interludes may be presented.

The system to be discussed will even allow a viewer to see new movies at reduced prices by selectively permitting advertising inserts in the subscriber's now less-expensive pay-per-view movie. This scheme could allow several price levels, depending on the total number of minutes of commercials the viewer is willing to tolerate. This, in turn, would allow the service provider to offset the reduced customer billing with advertising revenues.

The bottom line for the service provider should be: Which operating procedure, NVOD or TVOD, produces the largest revenue stream at what cost, ultimately providing the greatest return on investment? This chapter will summarize these issues.

System Possibilities

In order to analyze TVOD vs. NVOD costs, it is necessary to understand the three prominent hardware implementation philosophies illustrated in Figures 7.2 through 7.4. The differences between approaches depends on a vendor's reliance on his or her installed hardware architectures, as well as his or her philosophy on whether a general or "tuned" solution is preferable.

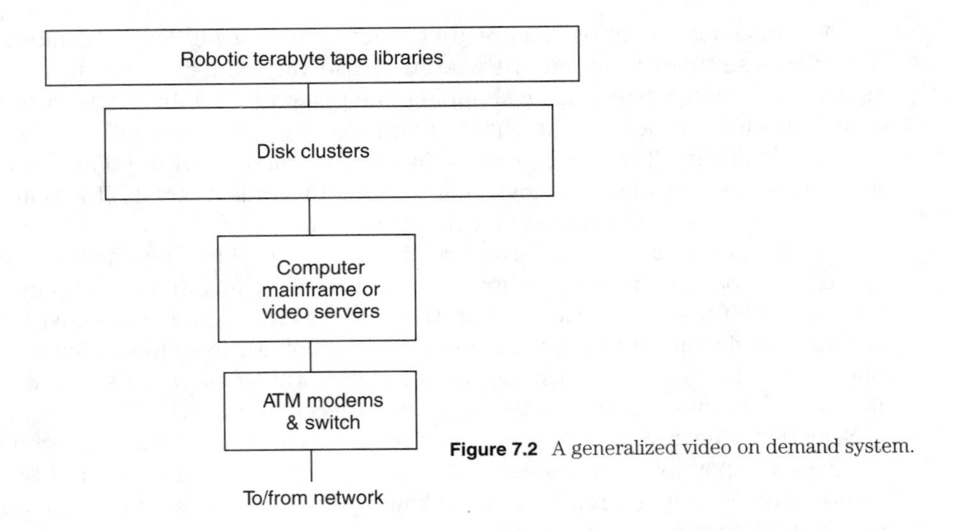

Figure 7.2 A generalized video on demand system.

In all the examples to be presented, it is assumed that the transmission system employs asynchronous transmission method (ATM). This protocol uses data packets consisting of a 5-byte header and a 48-byte data field. The header describes the destination and the content of the information portion of the packet. It is further assumed that the appropriate storage solution is a 3-level hierarchy of disk and robotically managed tape libraries. The general solution uses standard operating system functions and software, and the more "tuned" solutions employ significantly more specialized software and firmware to manage the hierarchy.

For applications where the volume is not adequate to justify a custom or tuned-design philosophy (such as for a small number of test sites, or for concept validation where reduced nonrecurring costs are important), the generalized solution as shown in Figure 7-2 may be preferable to a tuned solution. It is less expensive because it relies mostly on the procurement of off-the-shelf hardware and possibly off-the-shelf software. The generalized system can produce both TVOD and NVOD, but the cost of delivery is high.

In Figure 7.2, the term "mainframe" is intended to mean a general-purpose processor running a "standard" operating system (e.g., an RS6000 running UNIX). Such mainframe system solutions are often more expensive than tuned solutions in production, because a great deal of system hardware and software must be provided that is unnecessary for the specific application. Further, the mainframe data flow is designed for data processing, not data movement. Video applications require a great deal of data movement, with very little data processing.

The image processing (such as image compression and decompression) is usually performed by specialized hardware units. This is because affordable mainframes cannot handle the computational load required to deliver multiple video programs in real time.

When the opportunity exists to construct thousands of units for a specific application, the tuned solution is preferable because of lower cost, higher performance, superior function, and simply a better fit to the problem being solved.

There are various degrees of tuned systems. Some systems are very good at creating databases of still images or moving video that use general-purpose operating systems, database managers, networking facilities, and the like. These systems rely on small amounts of customization. They can do a good job of delivering a small number of selected videos on demand to a small customer base. As in the previous systems, they can produce either NVOD or TVOD, but the program selection is limited and the size of the client base is severely restricted when operating in TVOD mode.

These systems may be cascaded to accommodate more videos and more clients. An example of such a cascaded system is illustrated in Figure 7.3.

The ultimate tuning of a video server exists when special paths are provided for moving digital video information. An architecture can be created that relaxes the throughput requirements on the computer performing the server function.

Once the server has interpreted the customer's video request, validated billing and program availability, confirmed that the requester at the customer premises is not restricted (a child requesting an X-rated movie, for example), and arranged for the short-term scheduling (seconds or minutes), the server computer submits the program material request and the electronic customer address to the server saver/ATM switching system. Then, for the balance of that transaction, the server has nothing more to do until the program is complete (for a typical movie this would be between 90 and 110 minutes). This system is shown in Figure 7.4.

The server saver subsystem permits the use of inexpensive components and simplifies data routing and manipulation while simplifying computational requirements to such an extent that a single high-performance PC such as a Pentium or a PowerPC can assume responsibility for a 500-program, 10,000-subscriber system.

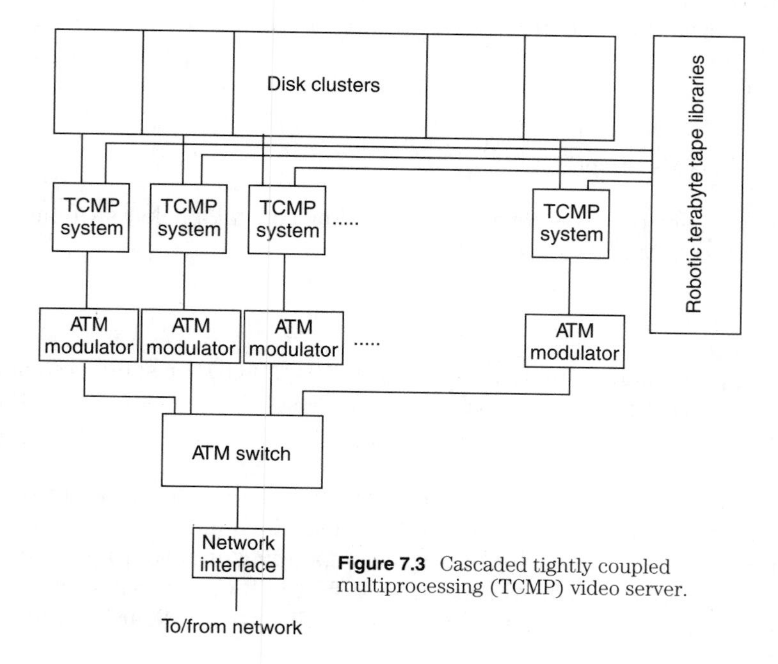

Figure 7.3 Cascaded tightly coupled multiprocessing (TCMP) video server.

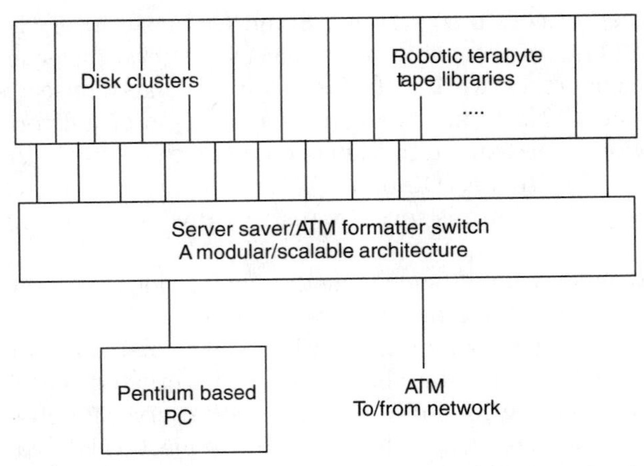

Figure 7.4 Composite server saver system/ATM switching system.

If a larger system is required, these systems can be cascaded to produce greater program selection for more patrons. The Server System can produce TVOD for a small number of subscribers or NVOD for a large number of subscribers, or some combination of both TVOD and NVOD. This capability is similar to that of the above systems, but *at very low relative cost.*

The server saver system is a simple device both architecturally and physically. It connects to a "storage farm" through multiple SCSI data paths, to the PC via one or more SCSI data paths, and to the CATV or other network through the ATM formatter/switch.

The server saver has only three types of interfaces:

- to/from the PC
- to/from the storage farm
- to/from the ATM network

The server saver provides storage control, flow control, packet switching, and an interface to the ATM network.

Costs

Each of the preceding systems can supply TVOD, NVOD, or some combination of NVOD and TVOD, but at substantially different costs. The cost of each of these systems varies as a function of program capacity, subscriber capacity, and the degree of responsiveness to customer requests.

It is obvious that video program capacity is a function of storage and that storage, in turn, makes up a major portion of system cost. Each 90- to 110-minute program can require from 1 to 9 gigabytes of storage, depending on resolution requirements. Each gigabyte of disk storage will cost from $750 to $1300 at the system level, while data in robotically controlled tape systems (e.g., StorageTek Nearline offerings) will cost $7 to $10 per gigabyte of storage. There are also performance differences between disk and such tape systems. These will be discussed later in some detail.

Obviously, the more programs desired, the more storage is required, which in turn increases hardware costs. The generalized video server systems typically cost $250,000 and upwards. Tightly coupled multiprocessing video server systems currently cost between $65,000 and $100,000 per module, each of which is capable of producing up to 25 programs concurrently. For example, 500 channels of programming could cost (500/25 × $65,000) or $1,300,000 per video server complex, not including ATM formatting, switching, or interfacing.

Each of these systems has limited capacity, requiring additional system hardware replication to yield more capacity and more responsiveness. Again, added system hardware increases system costs. The purpose of this chapter is to determine for the various generic hardware approaches the costs to produce the continuum between TVOD and NVOD, and how much responsiveness an interactive TV system can cost-effectively produce.

This chapter will generate some approximate best-case and worst-case pricing for each of this trio of approaches, determine reasonable pricing intervals, and evaluate the subsequent cost relationships for TVOD and NVOD. This will facilitate the qualitative judgment as to whether, for instance, it is worth an additional $500,000 or more to give the customer a program selection response time of 1 second/minute instead of 30 seconds/minutes. Further, after the analysis, procedures for camouflaging program latency will be discussed.

The following spreadsheet represents estimates of significant costs for each of the three prominent system architectural philosophies shown in Figures 7.2 to 7.4. While these numbers may be challenged as being tomorrow's prices, guesses, or inaccurate, they do represent working approximations derived from potential vendors in this industry. It is interesting to observe that using any set of different *reasonable* numbers does not change the comparative relationship: NVOD is much less expensive than TVOD.

This chapter has alluded to video programs and threads. A thread is defined as a continuous stream of video representing one complete program, using one of the available broadcast channels. Because both tape drives and video-friendly disks can produce data transmission rates greater than required for a single channel, it is possible to store the data in such a fashion that it can be read out multiple times in real time.

If a device is able to sustain a data rate 10 times greater than is required for normal video rates, 10 video streams or "threads" could be produced if only short-duration device-read interruptions occur (e.g., for turnaround at the end of a tape track, or for head or next-cylinder seeks). An alternative is for additional buffering to be used to mask longer-duration read interruptions. It is possible (and therefore desirable) to interleave the programming material so that each thread is displaced in time.

For example, a 90-minute (1 gigabyte) video program can be structured to allow 10 threads, and would have each thread offset by 9 minutes. This can be accommodated by appropriate data structures using only one gigabyte in either tape or disk storage. Because TVOD requires the ability to instantaneously access the first and then subsequent video frames of the program at random and arbitrary intervals, it would require that the storage device be capable of rapidly switching from one random spot to another to support even two threads, let alone a number as large as 10 or 12.

Although tape can support that many threads of NVOD, multiple-thread TVOD is not feasible with tape devices, because they require seconds to move from one random spot to another. TVOD is feasible but more expensive with disks because buffers must be included in front of each device for each thread, which substantially increases the cost per thread. More importantly, the randomly conducted seeks reduce the sustainable rate of the device so that it is less efficient, and even with external buffering it can sustain significantly fewer total threads.

For example: Assume a particular disk can sustain 3MB/s with a maximum (because the video stream must be guaranteed) random seek time of 33 ms. If the disk is rotated at 5400 rpm, it will have approximately 33K on a track that will spin by the head in 11 ms. (These, of course, are budgetary numbers, but may be adjusted for any particular device.)

If a random seek is allowed at the end of each track transfer in order to switch to another thread, then the sustainable rate is:

$$3\text{MB/sec} \times \{(11 \text{ ms/T})/(44 \text{ ms/(T+sk)}\} = .75\text{MB/sec}$$

If a video stream requires 1.5MB/s (~.2 MB/s), then the NVOD approach allows 15 threads without buffering, while the TVOD approach allows only 3 threads, *with buffering*. The buffer size, however, need not be very large, because:

$$.2\text{MB/sec} = .2\text{K/ms}$$

Then let buffer size for each thread = B_T

B_T = .2K/ms × (3 seeks of 30 ms + 2 transfers of 11 ms)

B_T = .2 × (90+22) K

B_T = 25K

Because each track must be buffered, it would adjust to 33K/thread = 100K buffer total.

This, of course, assumes the video-friendly type of device that has no other non-transfer activities to mask. Therefore, it is clear that TVOD threads with only one (or serendipitously, more than one) customer per thread will require many more disks than an NVOD with schedulable threads that allow a significantly greater number of customers per thread and a significantly greater number of threads per disk.

Furthermore, whereas the TVOD approach limits tape storage to 1 thread per device (as opposed to the 2 or 3 for the disk), the NVOD approach works as well from tape as it does from disk. The systems configured below are intended to support 1000 to 10,000 program titles, and use tape as the primary storage medium. The disks are used as a buffer for the currently active programs primarily to reduce the number of passes against each tape volume for reliability purposes, rather than for performance. As a matter of fact, tape performance in some instances will exceed that of disk devices in terms of the number of simultaneous threads that can be sustained. With NVOD threads scheduled in greater than 30-second increments (e.g., 5 to 15 minutes), the delay would completely mask the initial few seconds of startup to mount the tape.

Using tape directly, or using a disk as a buffer in front of the tape for most of the active programs (assuming the disk described above, and that each will hold 3 to 5 gigabytes), it would be possible to have each tape or disk provide as many as 15 threads (channels) of broadcast. This could be all for one program, or split among the number of programs that could be stored on that one device (e.g., three 90-minute movies would require 3 to 4½ gigabytes of storage). To support 200 channels of NVOD would require a minimum of 14 devices, and 500 channels would require a minimum of 34 devices.

The experience in this industry is that in any particular week there is a very small subset of programming that accounts for most of the demand. One specific example is for video rental, where 97% of revenue comes from less than 25 titles. With this tight a skew, out of a population of 1000 to 10,000 titles, between 33 and 68 titles account for 99% of the demand and between 39 and 129 titles account for 99.5% of the demand. See the inset below for the details on this set of calculations.

Customer Demand vs. System Performance: Limits Analysis

The given task is to identify the number of program titles necessary to satisfy a "large" proportion of the customer requests. Obviously the greater the percentage of requests one desires to satisfy, the larger the population. Also the distribution of the requests across the inventory of titles significantly affects the number requested. If the total number of titles is significantly greater than what can be simultaneously broadcast (e.g., more than an order of magnitude, such as 200 channels for 2000 titles), then the true answer will generally lie between an exponential and a hyperbolic distribution. Experience has shown that the number will quite often track hyperbolic through some significant portion of the range (e.g., 95% to 99%, depending on the tightness of the skew), then drift to the exponential, and then terminate at some finite number far short of where either distribution would predict.

Without knowing the actual distribution of requests to the most popular titles, it is difficult to calculate the exact number of titles that must be broadcast with any confidence. However, using what little is known about the reference patterns of the video rental base (e.g., one company reports that 97% of revenues come from 20 to 25 titles) one can calculate a range and bound the problem using distributions that historically tend to fit skew problems of this sort: i.e., binomial (to give an easy but very gross and optimistic first approximation), exponential, and "hyperbolic" (or "pareto") probability distributions.

The emphasis should be on the use of hyperbolic distributions (with the probability density form $p(n) = A/n^k$ for $n \geq 1$). It is a convention to use the word "3-sigma" to mean the value of the tail beyond $z = \pm 3d$ limits for the case of a "normal" or gaussian distribution (even though the actual distribution is not normal and may not even have a "sigma"). Framing the given problem between EXP and HYP limits gives the approximate value calculated here. One caveat is that historical skew distributions tend to deviate from perfect hyperbolic shapes at the high end tails (i.e., they drop faster than 1/nk and this is formally called "droop"). This shortens the real use tail so that the actual expected answer should be below that calculated at the 3-sigma limit for the hyperbolic distribution.

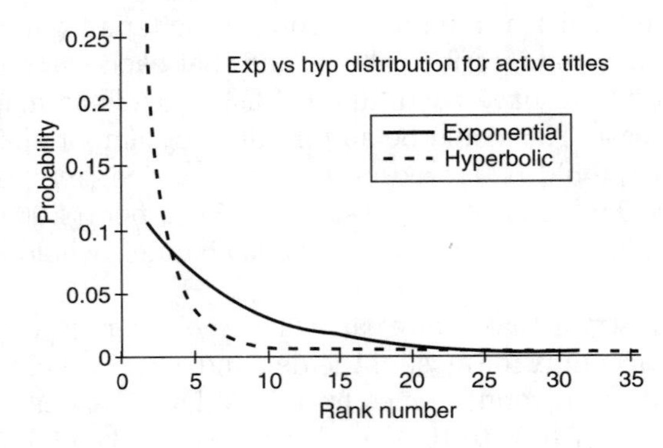

Figure 7.5 Exponential vs. hyperbolic distribution for active titles.

P = probability of selecting "choice" movie $\approx \frac{25}{10^F} = 0.0025 = \frac{C}{N}$

$\mu = Np = 25$

σ binomial % $\sqrt{Np} = 5,\ 3\sigma = 15$

$\mu + 3\sigma = 40$ Titles for 99.74% of demand

Same for Poisson (some N large, n small), $\mu = \lambda,\ \sigma = \sqrt{\lambda} = \sqrt{\mu}$, $\mu + 3\sigma = 40 \leftarrow$

Figure 7.6 Most elementary approach (binomial).

$$\int_0^{25} \alpha e^{-\alpha} dx = 97\% = 1 - e^{-\alpha 2\sigma}$$

so $\alpha = 0.14$

find n $\ni$ tail is 0.26% (Gaussian tail interpetation, 2 sides at 3α)

so CUM $= 1 - e^{-14n} = 0.9974$

$e^{-14n} = 0.0026$

$n = \underline{42.5}$

Figure 7.7 Exponential approach.

% Demand satisfied	25 Titles at 97% skew	25 Titles at 95% skew	20 Titles at 97% skew	20 Titles at 95% skew
99.74%	43	50	34	40
99.50%	39	44	30	35
99.00%	33	38	26	31

Figure 7.8 Demand vs. number of titles.

If the 25 titles were placed on shared disks at 12 threads each, and the rest of the programming spread over a few tape drives, the 200 channels could be supported by 12 disk drives and 12 tape drives. The 500 channels would require about 24 disk drives and 16 tape drives. The following cost analysis is on the basis of about 20 disks at 10 threads per disk plus 16 tape drives.

Individual disks and RAID (Redundant Array of Independent Disks) systems have different performance characteristics, so the numbers derived for individual disks and RAID systems is different. However, even when using the more expensive RAID technology, only a few TVOD threads can be produced.

$$\text{Assume } f(x) = \frac{A}{x^k}, \text{k unknown} > 1$$

$$\text{normalize (find A): } \int_1^{25} f(x)\,dx = -\frac{A}{k-1}\left(\frac{1}{N^{k-1}} - 1\right) = 100\%$$

our N very large => A~(k−1).

$$\text{Now CUM} = 97\% = \int_1^{25} \frac{k-1}{x^k}\,dx = \left(1 - \frac{1}{25^{k-1}}\right) \Rightarrow k - 1 = 1.09 \quad k \approx 2.089$$

(Note: $k \approx 1.931$ for 95% skew)

Then for 99.749.(30).

$$(k-1)1n(n) = -1n(0.26\%) = 1n(\text{"tail"})$$
$$n \approx 236$$

$$\text{e.g.} f(x) = \frac{1.089}{x^{2.089}}$$

Figure 7.9 Hyperbolic distribution.

% Demand satisfied	25 Titles at 97% skew	25 Titles at 95% skew	20 Titles at 97% skew	20 Titles at 95% skew
99.74%	236	599	161	385
99.50%	129	296	93	200
99.00%	68	107	51	100

Figure 7.10 Maximum values are presented. Actual values will be less.

		1 thread/disk Disk cost	10 thread/disk Disk cost	ATM Encoder/ ATM switch	1 Thread/disk system cost	10 Thread/disk system cost
COSTS for 25 thread video system						
MP server only	$65,000	$25,000	$2,500	$6,250	$96,250	$73,750
Server saver	$30,000					
Server saver+pentium	$40,000	$25,000	$2,500	$6,250	$71,250	$48,750
Mainframe	$250,000	$25,000	$2,500	$6,250	$281,250	$258,750
COSTS for 100 thread video system						
MP server only	$260,000	$100,000	$10,000	$25,000	$385,000	$295,000
Server saver	$120,000					
Server saver+pentium	$130,000	$100,000	$10,000	$25,000	$255,000	$165,000
Mainframe	$500,000	$100,000	$10,000	$25,000	$625,000	$535,000
Cost for 250 thread video system						
MP server only	$650,000	$250,000	$25,000	$62,500	$962,500	$737,500
Server saver	$232,500					
Server saver+pentium	$242,500	$250,000	$25,000	$62,500	$555,000	$330,000
Mainframe	$1,250,000	$250,000	$25,000	$62,500	$1,562,500	$1,337,500
Cost for 500 thread video system						
MP server only	$1,300,000	$500,000	$50,000	$125,000	$1,925,000	$1,475,000
Server saver	$480,000					
Server saver+pentium	$490,000	$500,000	$50,000	$125,000	$1,115,000	$665,000
Mainframe	$2,500,000	$500,000	$50,000	$125,000	$3,125,000	$2,675,000
Cost for 1000 thread video system						
MP server only	$3,250,000	$1,000,000	$100,000	$250,000	$4,500,000	$3,600,000
Server saver	$930,000					
Server saver+pentium	$940,000	$1,000,000	$100,000	$250,000	$2,190,000	$1,290,000
Mainframe	$5,000,000	$1,000,000	$100,000	$250,000	$6,250,000	$5,350,000

Figure 7.11 Assumptions in comparing the three VOD video server architectures.

Total# thread system	MP Server only system	Svr+srvr saver	Mainframe
25	$96,250	$71,250	$281,250
100	$385,000	$255,000	$625,000
250	$962,500	$555,000	$1,562,500
500	$1,925,000	$1,115,000	$3,125,000
1000	$4,500,000	$2,190,000	$6,250,000
	10 threads / disk		
25	$73,750	$48,750	$258,750
100	$295,000	$165,000	$535,000
250	$737,500	$330,000	$1,337,500
500	$1,475,000	$665,000	$2,675,000
1000	$3,600,000	$1,290,000	$5,350,000
	Cost comparison of 1 thread vs 10 thread systems		
25	1.305084746	1.461538462	1.08695652
100	1.305084746	1.545454545	1.1682243
250	1.305084746	1.681818182	1.1682243
500	1.305084746	1.676691729	1.1682243
1000	1.25	1.697674419	1.1682243

Figure 7.12 Table representing completed video server costs for a multiprocessor system, a server saver system, and a mainframe system. The last 5 rows of numbers represent cost improvement multipliers per thread.

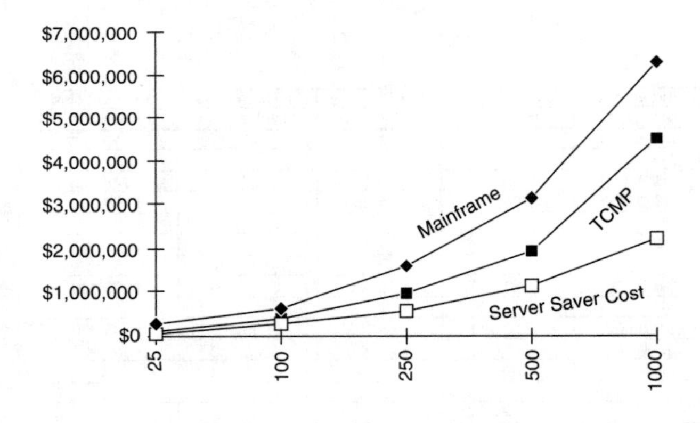

Figure 7.13 Chart depicting relative system costs for each of the 3 candidate TVOD video server system implementations.

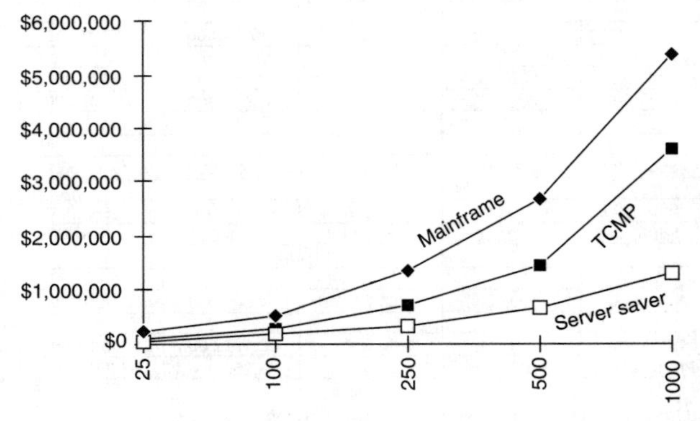

Figure 7.14 This chart represents NVOD cost per program thread for each of the 3 candidate systems assuming 10 threads are available from each storage device simultaneously. Depending on desired video quality and device performance, these numbers can change, but their relationships remain the same.

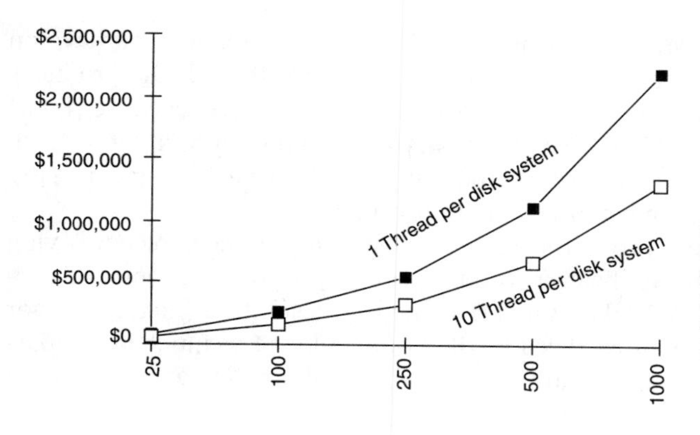

Figure 7.15 This chart illustrates the cost of the server saver application. The upper curve represents system cost when only 1 thread per storage device (TVOD) is provided, and the second curve represents system cost when a 10-thread-per-disk system is implemented.

The chart depicted in Figure 7.16 illustrates the cost savings as a percentage savings using the Server Saver System Architecture for 1 thread per storage device giving TVOD and 10 threads per storage device rendering unlimited capacity NVOD with a response time of 10 minutes. When the system program capacity is 20 units, NVOD can be produced for about 68% of the cost of TVOD, while systems above 250 programs flatten out such that NVOD costs less than 60% of TVOD systems, as depicted in Figure 7.16.

Figure 7.16 demonstrates how the cost per thread is reduced as the number of threads is increased. The vertical axis represents the cost relationship between the server saver system with one thread per storage device (TVOD) and the same server system with 10 threads per storage device (NVOD). Ten threads per storage device implies that for a 90-minute movie, 10 equally spaced start times can exist, providing a new start time for the movie every 9 minutes.

The horizontal axis represents the number of threads (channels) available to subscribers. The multiple thread system assumes that the disk storage system is video-friendly. Unlike standard drives, video-friendly drives are designed to provide a worst-case data rate that will ensure highly predictable delivery of data, so that discontinuities in the audio/video data stream will not exist.

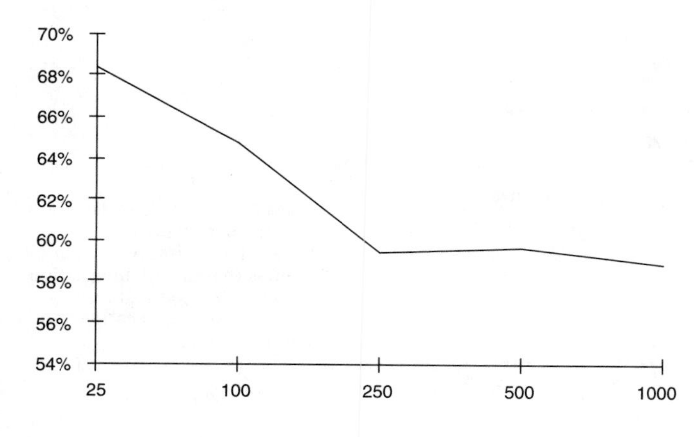

Figure 7.16 The above chart depicts cost savings of NVOD system over TVOD system for server saver-style architectures.

Figure 7.17 illustrates the savings that multiple thread disks (NVOD) can have on each of the candidate architectures versus single thread (TVOD). NVOD produces more programming at less cost per program than TVOD. Furthermore, because NVOD has an interval of time during which subscribers can request a program, NVOD can accommodate unlimited subscribers without requiring a subscriber to tune in late. Therefore, NVOD can produce substantially more revenue.

This chapter is also intended to determine the cost consequences of employing tuned solutions to the TVOD application versus general-purpose solutions or partially tuned solutions. Figure 7.18 illustrates that the tuned solution (i.e., the server saver architecture) with 200 or more threads will cost about 50% as much as the partially tuned solution (tightly coupled multiprocessor) and about 25% as much as the general purpose (mainframe) approach!

The crucial element to facilitate both TVOD and NVOD is the smooth, high-bandwidth, uninterrupted device transfer capability that we will refer to as "video-friendly." Interruptions in data will produce interrupted video, unless extensive and costly video buffering is provided. Interrupted video, of course, is unacceptable. It should be observed that only one vendor drive achieves this requirement (Micropolis) and one approaches this requirement.

Why use video-friendly devices

Video-friendly devices are able to cost-effectively produce multiple threads of smooth program video without the requirement for external video buffering, which requires extensive amounts of video RAM. Therefore, video-friendly drives are a significant component in reducing system cost.

Why use server saver-style architecture

The server saver architecture represents a highly tuned VOD application, not a generalized solution. It is the most cost-effective tool to solve the VOD problem. It provides significant advantages, including cost/performance, system flexibility, simplicity, high uptime, and low maintenance, as discussed in the author's previously referenced

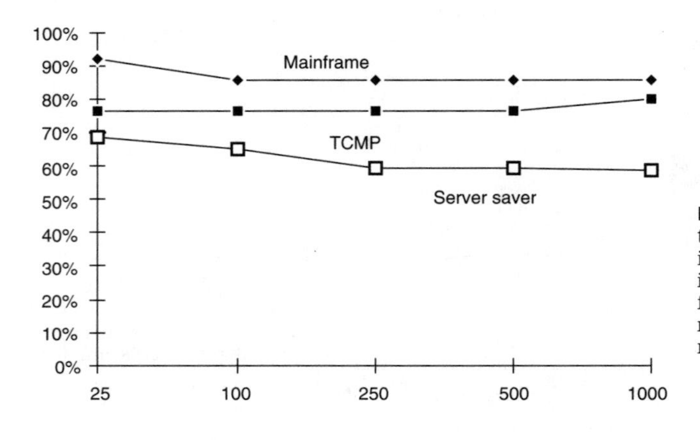

Figure 7.17 This chart depicts the same server saver information as Figure 7.16, but it includes the related information for the tightly coupled multiprocessor application and mainframe application.

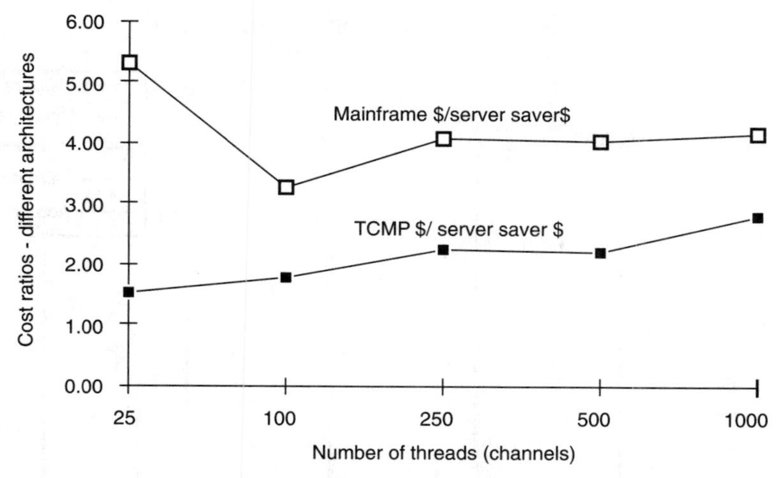

Figure 7.18 This chart depicts NVOD-based system costs normalized to the server saver architectural approach.

VOD articles (1). It can produce TVOD, NVOD, and combinations of TVOD and NVOD. Also, NVOD systems can produce unlimited customer showings per movie (unlike TVOD systems).

Why use NVOD instead of TVOD

NVOD systems can require approximately ½ the hardware cost to produce 10 times the video flow as do TVOD systems. Therefore, NVOD systems are the highly preferred economic approach. NVOD has been shown to cost substantially less to implement than TVOD and has the ability to support unlimited clients. NVOD can be tuned by the system operator to produce waiting intervals other than discussed in this chapter.

Perhaps an average 3-minute wait for the program is too long, even though that time is used for information on upcoming attractions, to sell food to be delivered to the home, to sell other services, or to merely provide a music-video interlude, or some combination of these. The system operator can reduce the NVOD interval by 50% while increasing hardware costs substantially less than 50%, thus moving closer to the TVOD model. This procedure can be repeated as often as desired to further reduce viewer latency time.

Studies in one TVOD vs. NVOD trial by a hotel pay-per-view TV operator indicated no increased revenue stream for the TVOD application, only added cost to provide the function to the same number of clients. One could make a career of looking at numerous other variations of data in the spreadsheet and graphing and plotting them. It seems obvious to the authors if an operator is decided on a TVOD system, he or she can use the server saver technology and video-friendly disks. If he or she desires the economies of NVOD, he or she can also cost-effectively employ the server saver technology.

If the operator is unsure of whether he or she wants TVOD or NVOD, he or she can use the server saver technology and provide both styles of programming to his or her clients. Statistics collected from the real world will probably tell the real story.

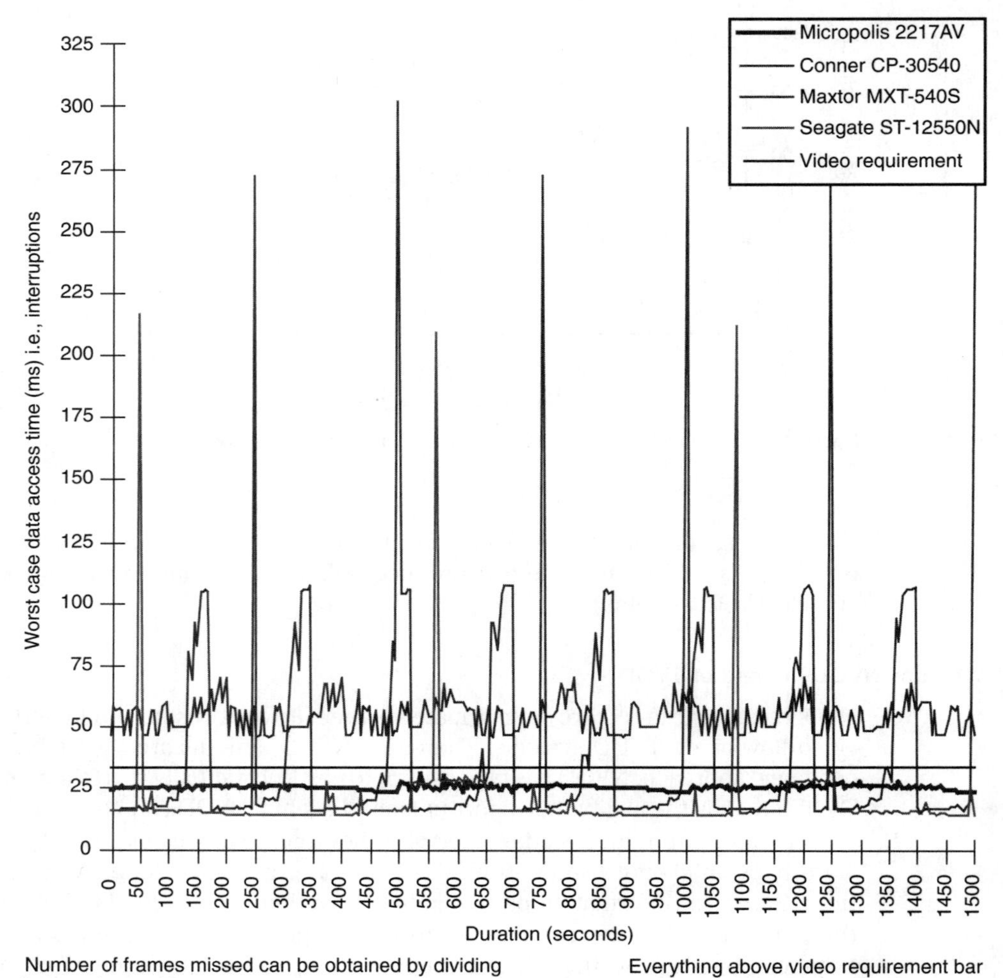

Figure 7.19 Transfer time vs. wall clock time for four representative vendor disk drives. The horizontal line at 33 ms represents the threshold of intolerable disk access times.

What is the impact of VOD on CATV-delivered ATM

The basic noncascaded server saver supports a 500-thread (or channel) system. The industry seems to support the idea of employing 50 MHz to 500 MHz for conventional analog TV and 500 Hz up to 1000 MHz for digital interactive TV, while leaving 5 MHz to 50 MHz for reverse channel communications.

If this is the case, then it is expected that as many as 500 streams or channels of digital interactive video could be placed in the upper CATV frequency band. If it were desired to support more program sources or threads, a different delivery system (such as fiber-optic cables) might need to be in place.

Because fiber-optic cable would only go to a city section, block, or curb, costly ATM switches would be required to move the proper packets from one transmission facility to another. This leads to the hotly debated question: Does a city require more than 500 channels of interactive TV, and if so, how much more will it cost to provide them?

NVOD will not require as many channels for transmission as TVOD to support the same number of viewers; hence it provides a great deal of relief from the expenses required to provide the infrastructure to support the greater number required by TVOD.

Filling in the Viewer Latency Time

The following strategy is proposed as a means of preventing the viewer from becoming frustrated at the delay between the time he or she makes his selection and the time it is actually delivered. Assume a maximum viewer latency time of 10 minutes. A number of prepackaged "mini-programs" may be prepared. They could be binary divisions of the 10-minute maximum time to be filled if a viewer requested a program only 1 second after the previous start time.

Thus, there could be one of several 10-minute cartoons, 5 minutes of coming attractions, 2½ minutes of news headlines, 1 minute and 45 seconds worth of public service announcements, 50 seconds of helpful hints, 25 seconds of quotable quotes, 12 seconds of inspirational messages, and up to 12 seconds of a warning that the feature is about to begin. Using various combinations of the above, any amount of time

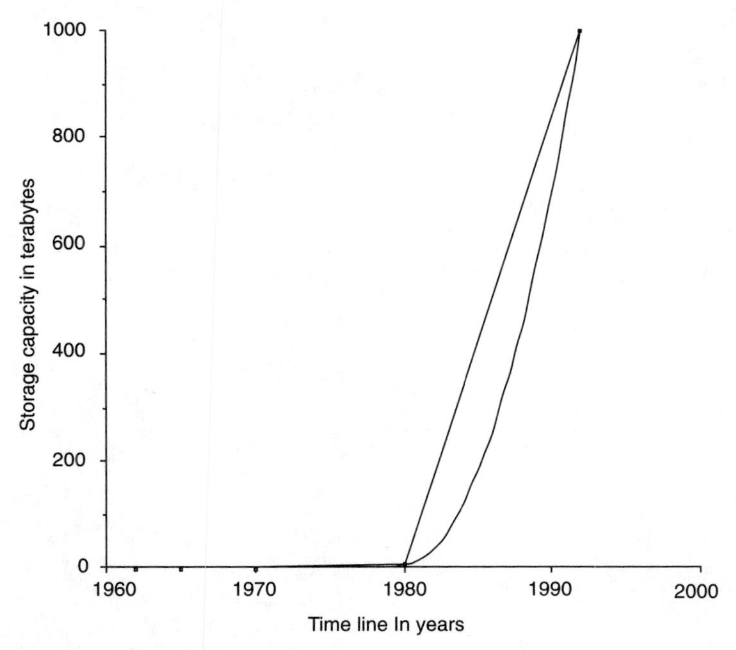

Figure 7.20 Transfer size vs. time.

up to the maximum latency time may be filled with entertainment. When it is determined what the delay will be, the viewer could be advised of the time remaining before the next feature starts, and given a menu from which he or she could select his or her own "fill in" entertainment.

Video-friendly disk drives provide almost a linear relationship between transfer length and transfer size. The data in this graph includes command overhead and is measured with the demand rate of 2.9MB/s. This figure is presented courtesy of Micropolis Corporation.

Conclusions

The ultimate goal of interactive TV is to provide the subscriber nearly instantaneous access to the programming of his or her choice. While this goal is attainable at very high cost, for a very limited number of subscribers, the authors do not believe it to be economically feasible to provide this type of service to the number of interactive TV subscribers projected over the next five years by leading industry market researchers.

NVOD offers a reasonable compromise between the ideal (zero viewer latency time) and an acceptable delay. This approach permits operators to obtain equipment that may be amortized by charges acceptable to subscribers.

Chapter

8

Storage Systems and Video Servers

Storage subsystems ranked in descending order by cost include RAM, disk, and tape. A hierarchical video storage server will likely be equipped with all three, with as little expensive RAM as possible.

The most important part of advanced video networks is the server. Configurations vary widely. Oracle's system employs nCube's massively parallel video server for its project with Bell Atlantic. Microsoft has developed a system that connects a number of personal computers to provide scalable video capabilities. Since these projects have commenced, a debate has raged over the fundamental architectural constructs of the servers themselves, and the costs associated with delivering digitally served video.

Next to the actual network infrastructure itself, video servers are the most capital-intensive piece of apparatus in interactive multimedia networks. Oracle claims that it can deliver up to 30,000 simultaneous streams of video for about $500 per household. Their price estimate does not include storage, which normally makes up about half of the server cost. Many analysts contend that at current levels, video can be delivered at about $1000 per video stream. However, the quality level of the video is yet to be determined.

Microsoft estimates that it can deliver video server capabilities at about one-tenth the cost of traditional servers on the market today. Microsoft's strategy is to start small and add PCs to the cluster as performance demands increase. This allows the computer hardware costs to remain more in step with the way subscribers will begin to use video server capacity—that is, gradually.

However, fundamental problems exist with Microsoft's approach to serving video. Media servers, unlike their data counterparts, require transmission of continuous, unbroken streams of information. This places a severe strain on the throughput capabilities of any system. An efficient video server architecture must be optimized for smooth video and audio through the entire path, from its source on the disk to its placement as ATM packets on the network. The video server must be optimized for smooth and efficient data movement.

It is a popular belief that video is more efficiently accessed when it is striped across several different drives. In other words, small packets of information are spread across several drives, maximizing the availability of any one drive. These drives are connected by bus structures that may possess varying throughput capabilities. Low-end PCs can deliver video at less than 5MB per second. This is insufficient for effective multiuser availability. On the higher end, Silicon Graphics claims its Challenger XL series (which is used in Time-Warner's Full Service Network) can deliver roughly 1GB per second through its bus.

This chapter will explain the benefits and architecture of a tuned video server architecture.

Technology and Economics

Media servers have to be powerful enough to provide reliable video-friendly service, cost-consistent with market demands. Every 18 months the processing capabilities double, while their cost is cut in half. Allocating one or more microprocessors to each customer using the network, as nCube and others have done in their servers, is low in cost relative to the overall price of server delivery. Memory and storage, however, make up the lion's share of video on demand delivery costs.

An uncompressed two-hour movie can consume roughly 100GB of storage space, while films such as *Jurassic Park* can occupy up to 150GB. The key to making the cost of storage and transmission viable is the use of compression schemes that can shrink the files down to between 1.5 and 8GB per movie (see chapter 5), providing compression factors and corresponding cost reduction factors ranging from 60 to 100 times.

Because of these memory-intensive movies and varying consumer demands for them, a variety of storage mechanisms must be employed. The most common form of server storage is the hard disk, which over the years has seen significant price drops. At about $0.70 per megabyte, a 1.5GB movie would require $1050 worth of disk storage space. Older movies and television shows will probably not experience the user demand necessary to justify their placement on the video server; however, they may still have a market.

To satisfy the lower demand for less popular videos, a secondary storage mechanism is essential (see chapter 7, TVOD/NVOD). Robotic tape libraries can be used to fetch infrequently viewed movies from an inexpensive archive and present them to a limited viewing audience.

Due to their high cost, semiconductor RAM storage mechanisms are impractical for video storage. At roughly $20 per megabyte, storing a 1.5GB movie would use more than $30,000 worth of RAM. RAM, however, is an essential part of a server system. It acts as a video and audio smoothing buffer that is capable of holding the necessary amounts of programming material while disks seek, recalibrate, and correct errors. Video-friendly disks require significantly less RAM storage, because they are designed to produce a smooth stream of audio and video data.

A Case for Near Video on Demand

"Will consumers support the cost of true video on demand?" This question was asked of more than one thousand attendees at the Video On Demand Conference at the 5/24/94 National Cable Television Association Convention in New Orleans. More than 80% of the audience expressed the belief that while the convenience of the immediate program selection of TVOD was clearly compelling, the cost renders its viability questionable. There are ways for video providers to provide apparent instantaneous interactivity and still amortize NVOD infrastructure costs to achieve a practical video network. Market testing has demonstrated that advanced networks costing more than $1000 per user cannot be economically justified.

But the tests indicate that demand *does* exist. They also act as an endorsement of a scalable equipment evolution. True video on demand requires several components if it is to be able to deliver video at the touch of a button and maintain full VCR-style control. System requirements include a switched, wide-band network and very high-powered servers equipped with many disks. In contrast, near video on demand service requires only increased channel capacity and a scalable server.

The true cost savings of NVOD comes from its independence from completely random user start times. The number of video streams required is dependent only on the length of the time increments for the programs. For example, a 90-minute movie with 30-second start times would require 180 video streams, one-minute increments would require 90 streams, and two-minute intervals would require 45 streams. Because each 1GB video-friendly disk can produce 15 streams, the corresponding scenarios would require 12 disks, 6 disks, and 3 disks, respectively. Scenarios 2 and 3 are clearly practical, while scenario 1's feasibility is questionable.

It would, of course, be absurd to consider lining up 240 VCRs to stagger start times for one movie, when video can be striped across several disks in concurrent time intervals. For example, one two-hour movie striped across seven disks could permit one-minute starting intervals as well as pseudo-VCR controls. In other words, viewers could jump backward or forward in the video in one-minute increments as much as they want, without requiring their own dedicated video stream. This is video stream sharing.

Perhaps the most intriguing attribute of the near video on demand scenario is its ability to generate revenue when movies are not even being shown. Unlike true video on demand, which offers instant access, near video on demand leaves windows open between the starting times of shows; these windows can be used to generate revenue through advertising.

Because NVOD staggers starting times at intervals ranging from 30 seconds to 10 minutes, the waiting interval offers the opportunities for retailers to advertise goods and services, thereby producing additional revenues. At the present time, delivery of served video remains severely limited by the low uncompressed channel capacities of most cable and telephone networks. However, the costs of infrastructure upgrades may be justified by additional revenue streams produced by servers.

The cost curve of video servers will continue to decline as new applications arise. Applications are being found in commercial settings, such as newsrooms for ad insertions, or storing video for editing. Cable News Network, for example, uses a server in its Atlanta studio that allows producers, editors, and writers to view video clips at their desks. Home Shopping Network is also developing a real-time server to enhance its own services as well as to sell to outside parties.

Several servers have already made their market debut and have seen steady growth rates. Machines from IBM, Hewlett-Packard, Silicon Graphics, AT&T, nCube, and Fujitsu are a few of the large players with servers to offer. Smaller companies such as The Network Connection, ProtoComm, and StarLight Networks have also marketed scalable server solutions for a variety of commercial server applications.

The paramount issue for any server is interoperability. The advent of interactive television is bringing a myriad of home management devices, network connections, and operating and application software to the market. Servers need to be technology-independent and set top box-friendly in order to keep networks open and support the multiple devices that will not only justify the *existence* of the server but also its *price*.

Video Friendliness

Common computer disk drives deliver data to computers as fast as possible, but they have no requirement to deliver it as smoothly as possible, because computers operate very well with irregular or jerky data transmission. Irregular data rates are due to random seeking, error correction, and recalibration. Jerky data produces a jerky video stream. This can be smoothed only with a great deal of expensive RAM buffer-

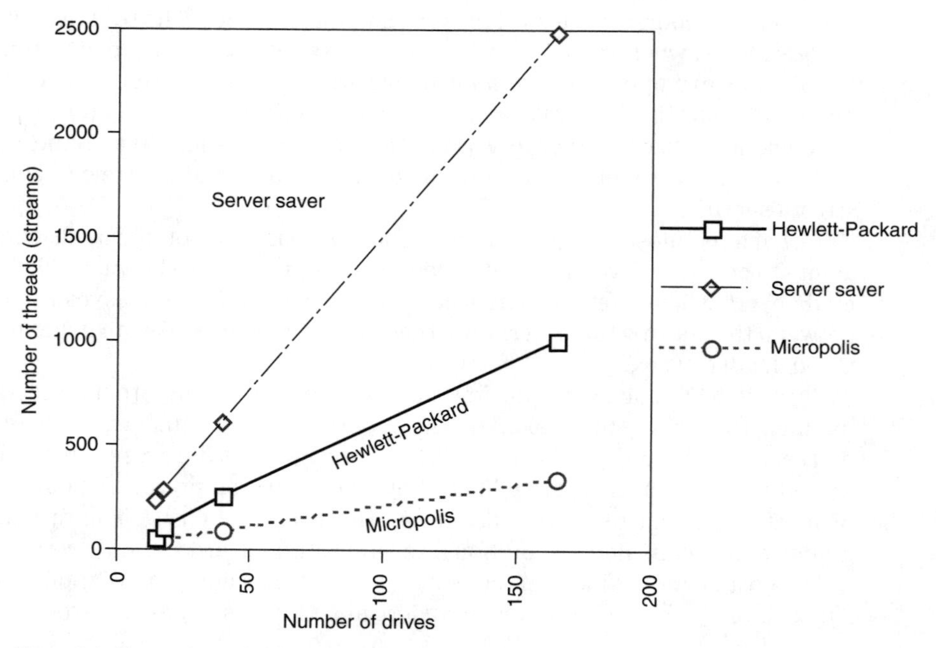

Figure 8.1 Comparison of video server threads per drive.

ing. Jerky data is considered to be video-unfriendly. Video-friendly drives permit smooth and continuous flow of data between the disk drive and the video server or multimedia computer.

The philosophy inherent in video-friendly disk architecture is that it is desirable to:

1. Recognize an error rather than stop the flow of video to correct it.
2. Take advantage of the inherent phenomenon of sequential seeking prevalent in video data.
3. Take advantage, wherever possible, of self-identifying embedded addresses in the video stream to make calibration unnecessary.
4. Take advantage of the disk transfer bandwidth to produce multiple streams of video from one disk.
5. Provide a simple command set optimized for video record and playback.

Depending upon available disk bandwidth and required video resolutions (which may range from VHS-quality to HDTV-quality), 3 to 18 streams of sequentially accessed video can be produced from one disk drive. In NVOD applications, this offers opportunities for significant system cost savings by relieving the system of the burden of extensive RAM buffers. Video-friendly disk drives are available from Micropolis and Conner Peripherals at current (7/94) costs of about $600 per gigabyte. These prices are expected to erode substantially with time.

The performance of virtually all styles of system architecture improves with video-friendly disk drives versus video-unfriendly disk drives. The video-friendly disk drive is an important ingredient in the video server and multimedia computer environments.

Video Server Architecture

As described in chapter 7, three basic video server architectures have been developed to facilitate video on demand. These implementations include:

- A general-purpose mainframe approach (Figure 8.2)
- A tightly coupled multiprocessor approach (Figure 8.3)
- A specifically designed multithread video server (Figure 8.4), also called "tuned video server" architecture.

The general-purpose mainframe approach to video serving implementation is expedient to implement, quite costly, and as reliable as the mainframe system permits it to be (Figure 8.2). The tightly coupled multiprocessor approach requires slightly more time to implement, handles more concurrent programs, and is less expensive than the mainframe implementation. Reliability is higher than the mainframe approach, because the greater redundancy (Figure 8.3) can be exploited.

The multithread video server implementation is tuned for the video serving application. It is modularly expandable to support as many parallel digital video data paths as desired or required. This is possible because a video server is primarily a data moving and routing system and does not require extensive computing capability.

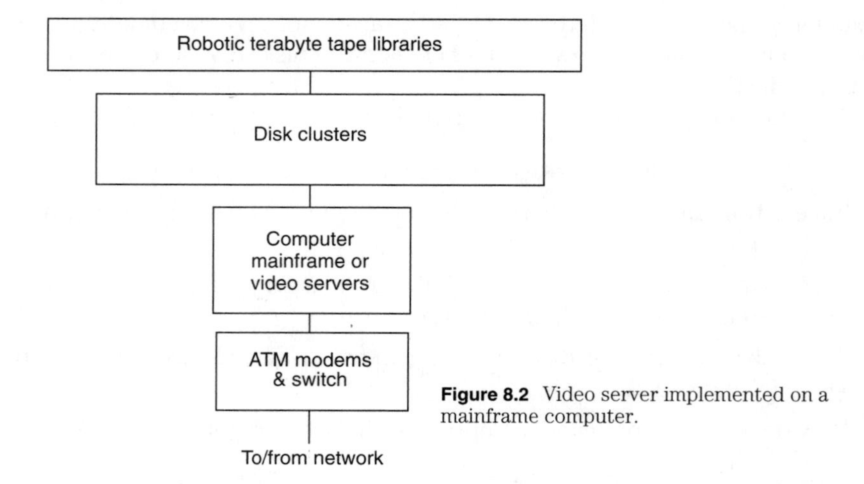

Figure 8.2 Video server implemented on a mainframe computer.

Each data path is able to support multiple video threads. Extensive redundancy exists facilitating significant reliability improvements over the two previous approaches. A special video server instruction set makes programming more efficient than on machines employing general-purpose instruction sets. Additional special functions facilitate easy access to the ATM network (Figure 8.4).

Architectural Comparisons

Given an understanding of the video server requirements, several ways to design the video server become obvious.

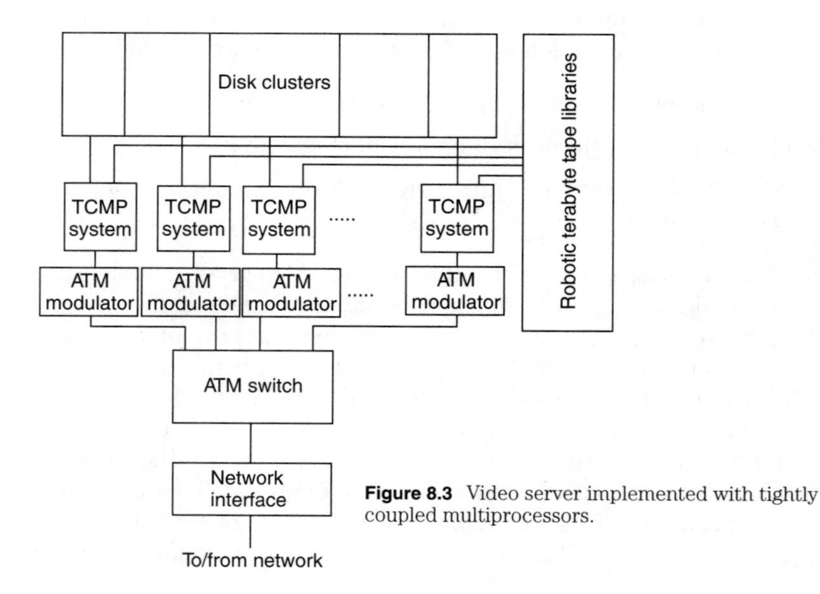

Figure 8.3 Video server implemented with tightly coupled multiprocessors.

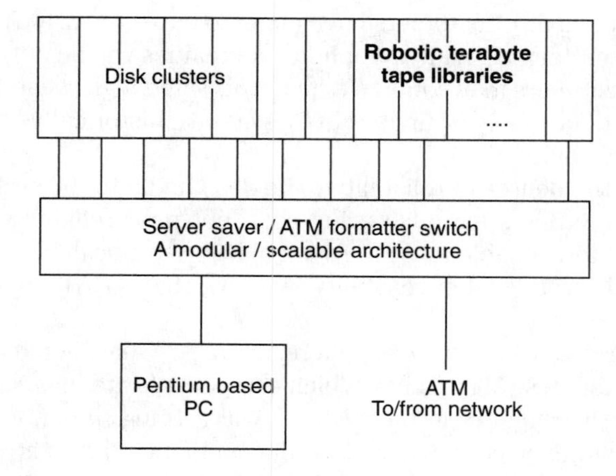

Figure 8.4 A specifically designed multithread video server (optimized video server).

The simplest approach, of course, is to implement the video server functions on a mainframe with enough performance to permit real-time operation (Figure 8.2). Alternatively, a system employing a tightly coupled multiprocessing architecture (Figure 8.3) may be programmed to execute the desired function. The last approach (Figure 8.4) employs a completely custom and finely honed design encompassing all the objectives and functions required of a video server.

A chart representing the relative "goodness" of the various approaches is shown below in Figure 8.5, where higher numbers indicate greater cost-efficiency of the design approach.

Reliability

The tuned (honed) video server is designed to provide the highest reliability for the lowest possible cost. However, different customers have different ideas of what is acceptable reliability. For instance, it is not uncommon for cable systems to go down several times a year. During such a failure, there may be several hours where all channels on the cable are interrupted. Many customers grumble about this, but are not dissatisfied enough to cancel the service (of course, there is no competitive service).

	Mainframe	Tightly coupled multiprocessor	Tuned video server
Reliability	good	very good	outstanding
Modularity	poor	very good	outstanding
Scalability	poor	very good	outstanding
Serviceability	good	very good	outstanding
Interactivity	good	very good	outstanding
Cost	poor	very good	outstanding

Figure 8.5 Comparison of video server approaches.

In contrast, telephone company customers have come to expect nearly 100% reliability of telephone service. Because many telephone companies will be entering the ITV marketplace, most customers may come to expect this same high reliability from their CATV service. Telephone systems, unlike CATV systems, historically have been designed for maximum reliability.

It is necessary to know the degree of reliability to be designed into the system. Reliability frequently directly affects manufacturing cost, but poor reliability creates customer dissatisfaction. How reliable must the system be? Can a scalable and modular system provide the desired level of reliability at a cost that a CATV operator is willing to pay?

From what kind of failures can a modular system recuperate? What should the minimum mean time between failures (MTBF) be? Which components are most apt to fail, and what is the effect of their failure on the system? Will it bring the entire system down, or just a single program, or perhaps just introduce a momentary "glitch" in the program? What happens if a second component fails before the first is replaced?

With appropriate design, redundancy improves system reliability. If redundancy is designed in, what level of reliability should be accommodated? Should there be completely redundant systems, with one simply shadowing the other, waiting to be switched in the event a failure is detected in the primary system? Should a failed subsystem be automatically switched out and replaced by a new one upon a failure detection, or should the system operator merely be notified so that a replacement module can be manually "hot swapped"? Is cold swapping acceptable? All these are questions that must be decided by the video supplier, based on his or her assessment of the cost/benefit trade-offs affecting his or her particular marketplace.

Redundant arrays of independent disks (RAID) systems for video provide an added disk drive on which is recorded error correction information. Typically, a 5- or 6-drive RAID system will have an added drive providing the necessary information to facilitate real-time error correction. If one disk fails hard, the system will continue to operate, but if a second drive fails, all the programs striped across the remaining group of drives will fail.

The tuned video server can be designed to provide the same expected failure protection as the RAID system for the first failure. However, it can tolerate a second failure with consequences ranging from a few second "glitch" to the loss of a program as stored on a single disk. Significant cost trade-offs can be made. For example, in the event of a failure resulting in the loss of some program threads, the system could automatically switch to a previous thread of the same program. The customer would then experience only a momentary inconvenience with perhaps as little as a 30-second repetition of a previously viewed program segment. This could offer a cost-effective alternative to complete redundancy of each disk system.

Modularity

System modularity should permit the cost-effective implementation of a video server system that handles as few as 10 programs and can be expanded to a system delivering thousands of programs or threads. The tuned video server could be configured as shown below: an electronic box supporting up to 50 disk drives and 750 threads.

Multiples of these boxes could be connectable to each other and to a multicable system via standard ATM switches.

The tuned video server can be directed by one or more personal computers that provide the control and billing functions. Such a system is shown in Figure 8.6.

Figure 8.7 illustrates the placement of function and the modularity of the tuned video server. To expand the system, one merely adds more disk controller cards. Each controller supports a full SCSI interface to each disk, but for added economy, multiple disks could reside on each SCSI port. To improve reliability, redundant SCSI ports could be connected to some number of disks arranged in strings. Such an arrangement would permit varying degrees of reliability as a function of cabling configuration and cost.

The highest-performance system is configured as shown in Figure 8.7. Here each SCSI controller is dedicated to one disk, permitting maximum throughput, and consequently the maximum number of threads. This is also the tuned configuration with the highest cost, but as shown in chapter 7 this architecture remains the lowest cost for the given reliability level.

The tuned video server architecture shown in Figure 8.8 shows how the disk controllers and ATM packetizer(s) are connected to the Smart System motherboard.

Because it may be desirable to have redundant billing and control computers for reliability, the connection to the control and billing computer should be a standard interface. This would permit multiple billing computers to be connected via SCSI,

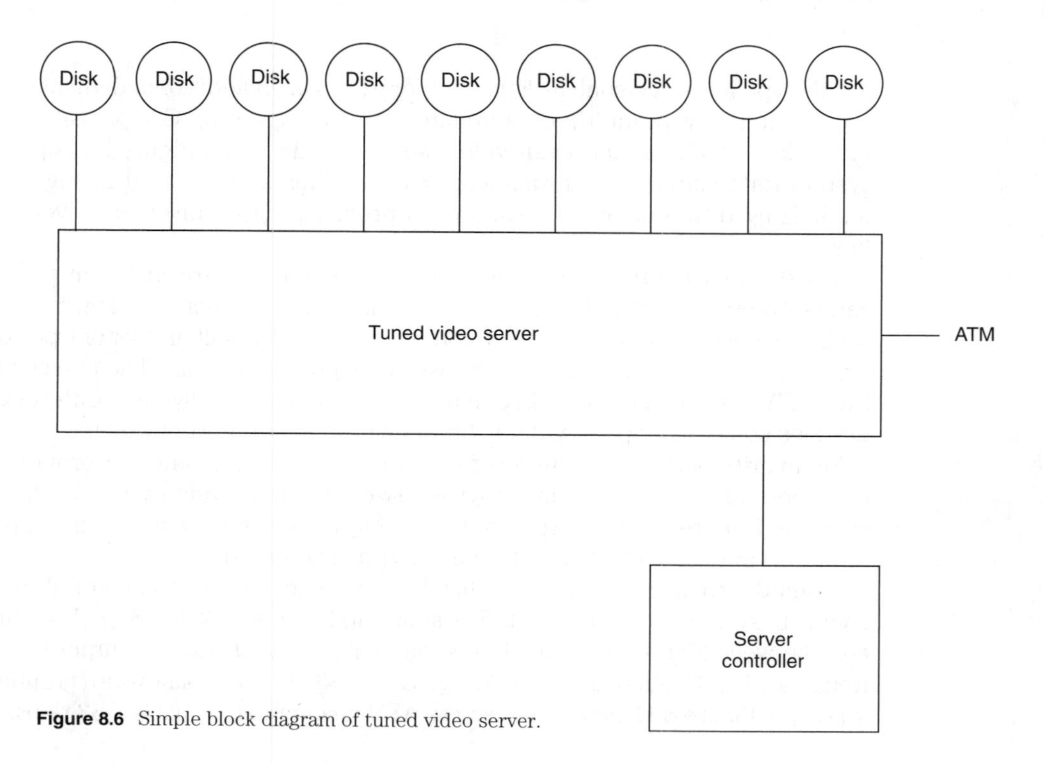

Figure 8.6 Simple block diagram of tuned video server.

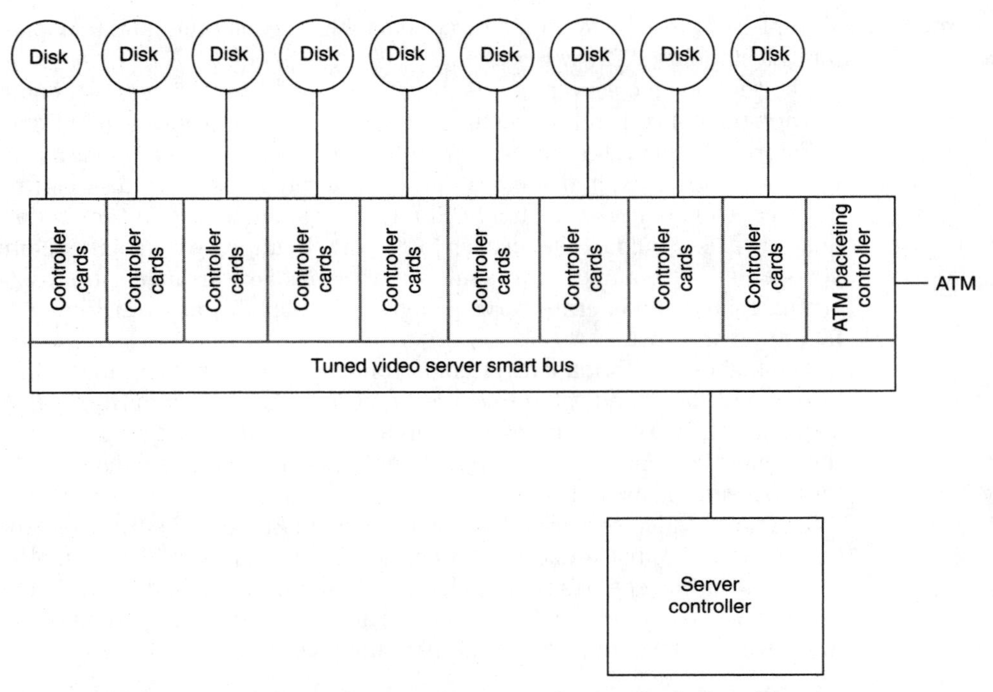

Figure 8.7 Expanded block diagram of tuned video server showing distribution of functions.

a LAN type interface such as Ethernet, or even via an additional dedicated ATM interface. In this way, multiple redundant billing computers, ATM packetizers, multiple disk controllers, and even video servers could be configured to optimize the system for reliability, performance, or costs. This is illustrated in Figure 8.9. By minimizing different part types, spares provisioning requirements would remain low.

The dynamics of the tuned video server system design are such that performance can be traded for reliability in real time. For example, when considering configurations for cost trade-offs, a failed component could result in system performance degradation, or at a slightly higher cost, no degradation at all. The choice would be the CATV system operator's. In a dynamically configurable system, different considerations could be emphasized for the same system at different times.

Modularity permits system reconfiguration for reliability and performance. Scalability permits reconfiguration for system size. The tuned video server architecture is scalable. If more disks are required by the system, the only requirement is to plug more server disk controllers into the smart motherboard.

When the smart motherboard is filled to its capacity of 50 disk controllers, another complete system can be connected as shown in Figures 8.10 and 8.11. The only difference between Figures 8.10 and 8.11 is that in Figure 8.10, a single output port delivers frequency bands of modulated ATM signals at DS3 or 6-MHz bandwidth groupings. The system of Figure 8.11 produces discrete ATM channels at probably DS3 rates.

Interactivity

The degree of viewer interactivity that the video server can support is limited by disk seek time (nominally 10 ms) and by program decision possibilities, rather than by the tuned video server. Using the tuned video server architecture, increased interactivity comes at the expense of supporting fewer users per thread. When longer sequences of video exist between decision points, more customers can share that video database. For example, going down the aisles of a grocery store, a navigational decision is required only at the ends of each aisle.

Serviceability

The scalable architecture uses only four different types of printed circuit boards. Each board is microprocessor-based and has ROM diagnostics and three display lamps spec-

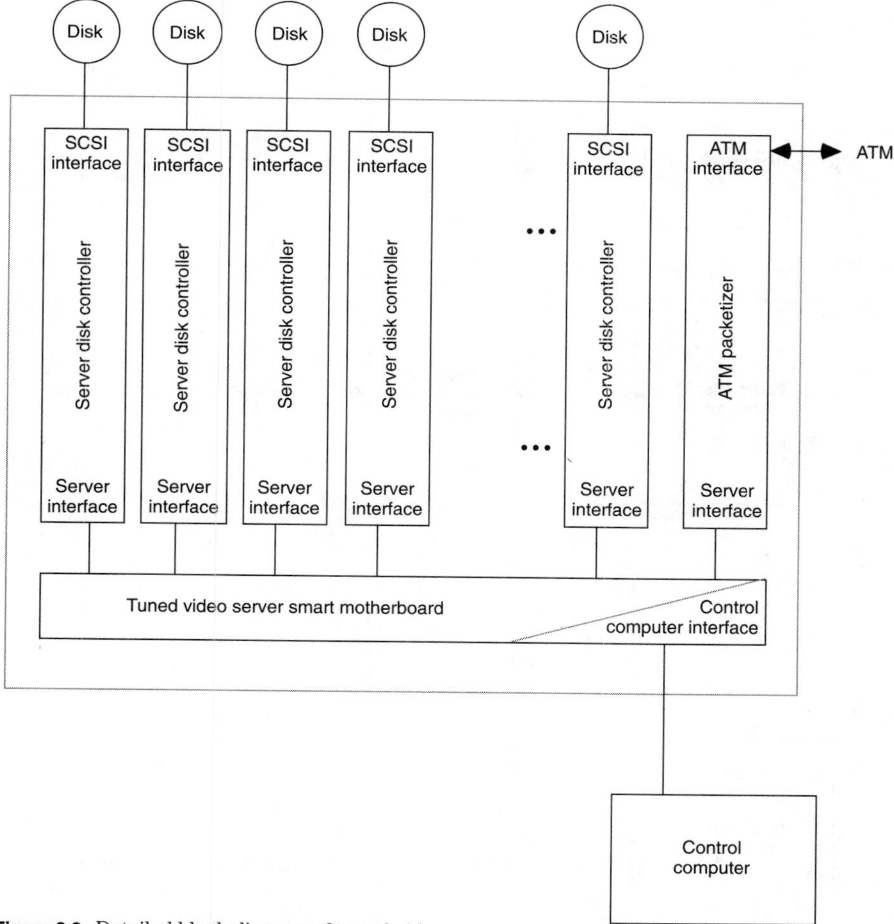

Figure 8.8 Detailed block diagram of tuned video server.

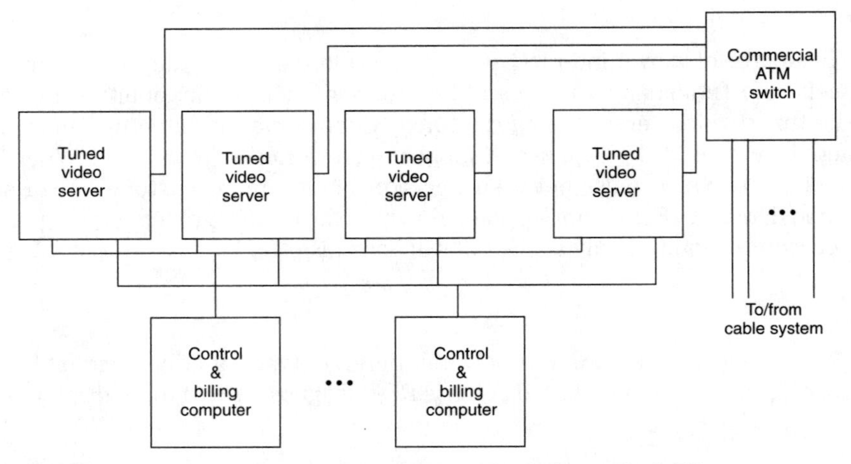

Figure 8.9 A highly redundant configuration providing bulletproof reliability.

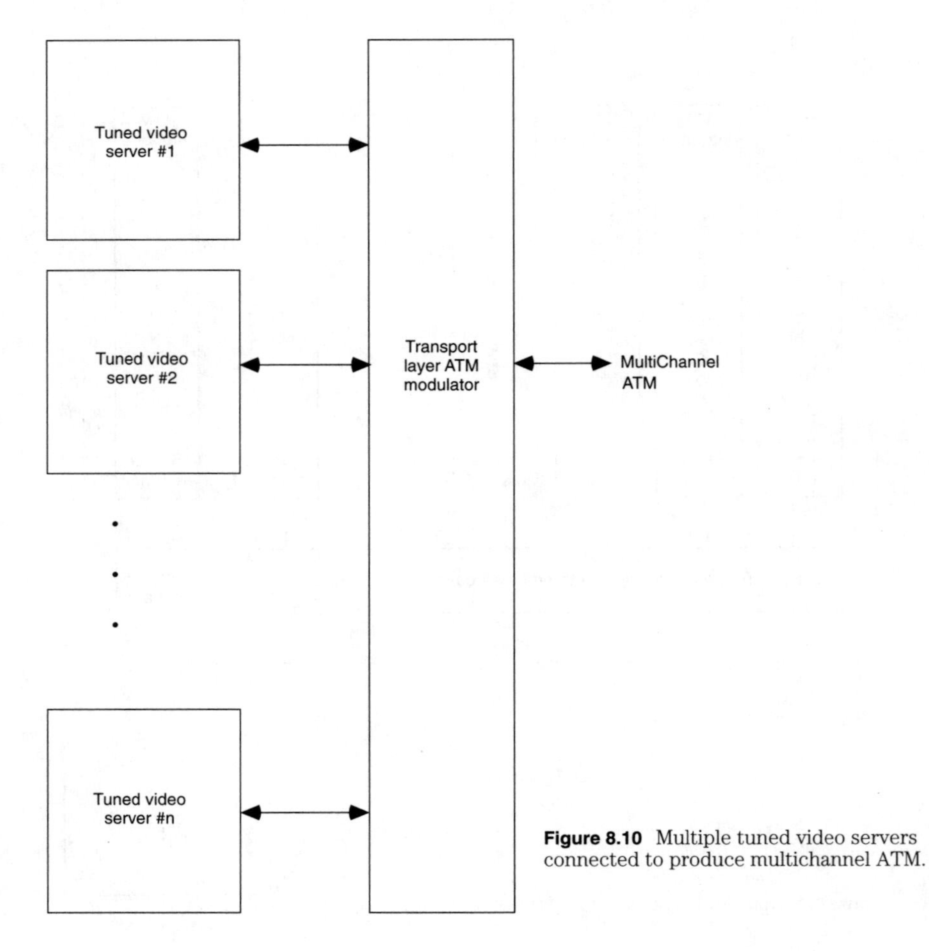

Figure 8.10 Multiple tuned video servers connected to produce multichannel ATM.

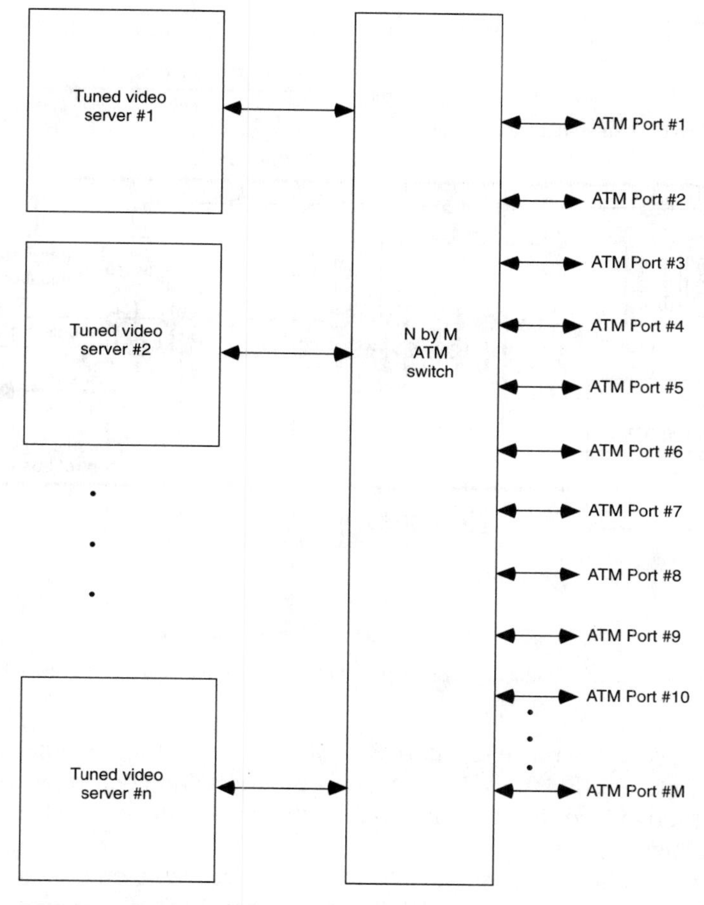

Figure 8.11 Multiple tuned video servers connected to produce separate ATM channels.

ifying the status and the nature of a failure, if one exists. Two buttons on each board permit either off-line or in-line diagnostics. In-line diagnostics run concurrently with normal system operation, collecting and displaying status information continuously.

In the event of a malfunction, status is immediately transmitted to the control and billing computer. The billing computer then provides control information to the video server describing how it should handle the fault. The billing computer also can (via modem) call a designated central status monitoring point to dump its status. This central point can dispatch repair personnel as required.

Instruction set

The tuned video server is designed around emulator-style architectural building blocks. These building block emulators each have about 20,000 gates (quite small for VLSI) and have wide microprogram control words providing many functions during

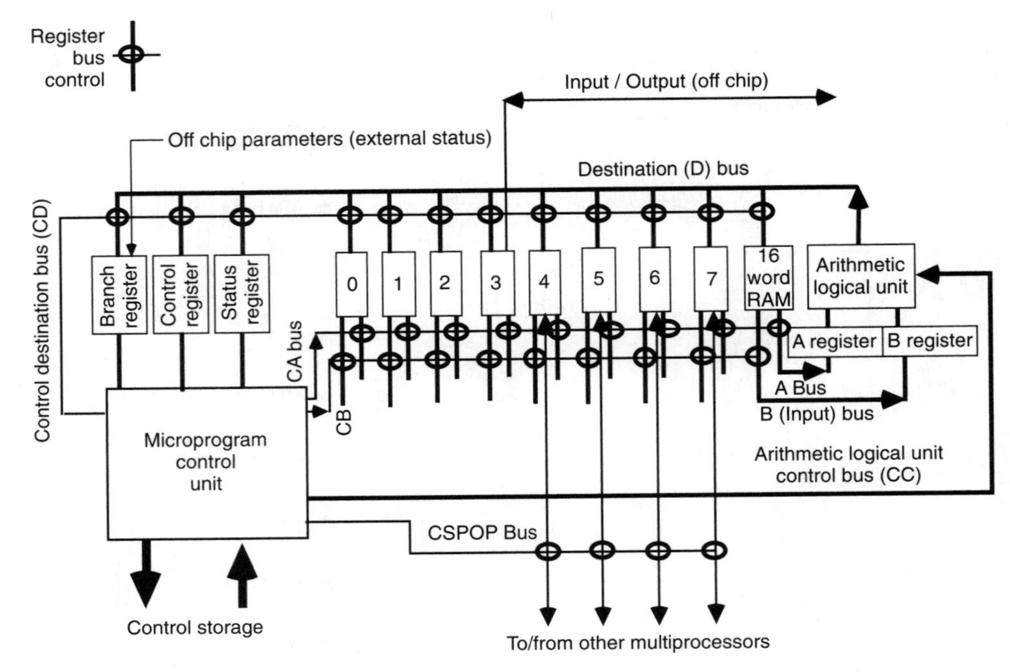

Figure 8.12 Typical emulator building block employed throughout tuned video server architecture.

every 35-nanosecond microcycle. Figure 8.12 depicts one of these emulator building blocks, while Figure 8.13 details the emulator's microprogram control word. The instruction set (and indeed the entire function of the tuned video server) is defined by groups of microcoded instructions.

The motherboard for the tuned video server is controlled by one of these embedded emulators with unique microcode. The disk controllers, ATM packetizers, and billing computer interface all employ these very-high-speed emulators to provide their functions and interfunction communications. The emulators are very tightly coupled via special registers, buses, and microcode.

The instruction set of the tuned video server as viewed by the control and billing computer consists of the following:

- Abort program #_______
- Change media program virtual address
- Change program's virtual address
- Change thread
- Collect subscriber "N" status
- Control set top box real address #______
- Control set top box virtual address #______
- Copy disk #_____ to disk #_____

- Copy tape #_____ to disk #_____
- Deselect media program
- Diagnose communications interface
- Diagnose media program
- Diagnose set top box #______
- Diagnose system
- Download control store to all set top boxes
- Download control store to set top box #________
- Insert new thread
- Pause program
- Query number of threads for selected program
- Read media program from specified disk
- Request set top box #______ status
- Reset
- Reset retry facility
- Search for media program
- Select media program
- Select media thread
- Select set top box by real address _____
- Select set top box by virtual address ______
- Select subscriber
- Selective disk reset
- Selective set top box reset
- Send new virtual address to set top box real address #_____
- Sense communications errors
- Sense disk errors
- Sense media interaction point
- Sense media program #_____ end of program
- Sense media program #_____ end of scene
- Sense pending end of media programs
- Sense set top box #____ viewing time
- Sense set top box errors
- Sense summary status
- Sense system errors
- Sense system errors

- Set media interaction branch possibilities
- Set reliability options
- Set retry facility
- Set set top box available functions
- Set set top box decoding procedures
- Set set top box restrictions
- Start selected program
- Start selected program at time ____
- Validate disk #_____
- Write media program to specified disk

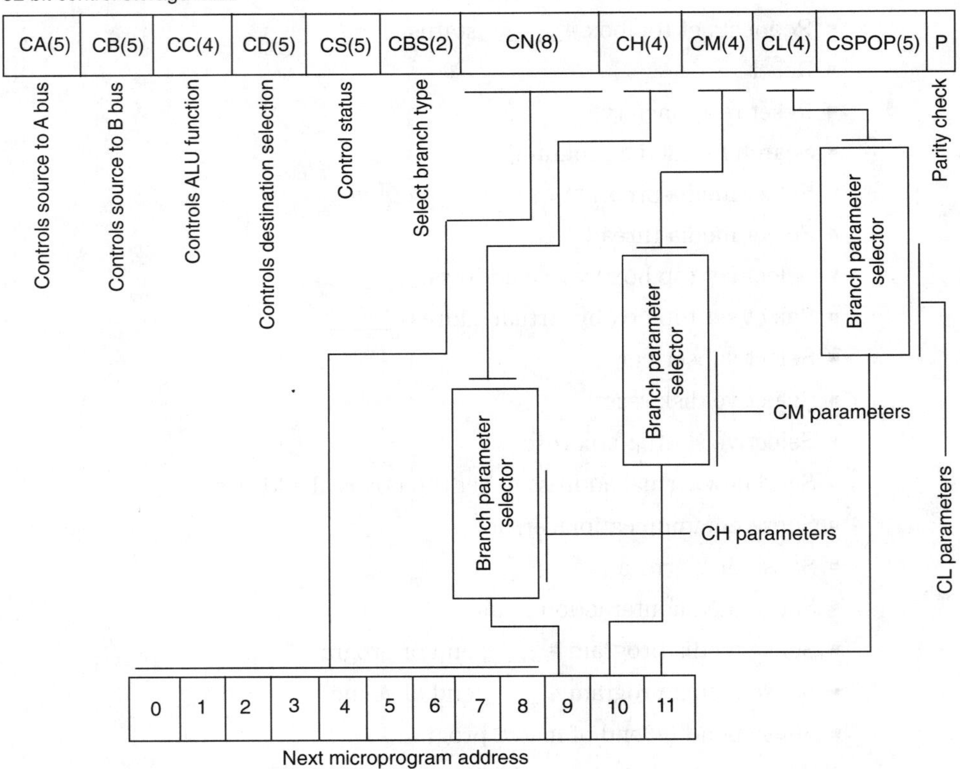

Figure 8.13 Control word for emulator building block shown in Figure 8.12.

Summary

Video server architectures range from mainframes implemented with massively parallel microprocessors, through scalable, tightly coupled high-end microprocessors, to systems specifically "tuned" for the video server application. The latter are characterized by the use of video-friendly storage devices and data paths optimized for the smooth data movement from disk to delivery media. These are required to provide "glitchless" video to the viewer.

The mainframe is the easiest to implement. The tightly coupled multiprocessor system is easier to scale to meet the changing volume requirements of the video provider than is the mainframe approach. However, the tuned solution is the most cost-effective solution to meeting the "cradle-to-grave" requirements of the station operator.

Reliability may be implemented to whatever degree the operator feels meets the expectations of his or her customers and the financial resources available to him or her. The open, modular approach permits a wide variety of configurations to provide the amount of redundancy required to meet his or her system availability goals.

An architecture implemented by a large number of identical, microprogrammable, embedded LSI modules provides a minimum of board types for minimum spares provisioning. It also provides for automatic reconfiguration upon the detection of hardware failures, and speeds the diagnosis and repair of failed components. The control word format, interconnection scheme, and instruction set of the proposed video server architecture were presented.

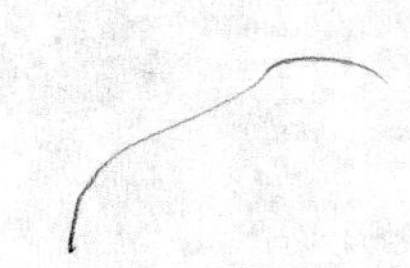

Chapter

9

Switching, Traffic Control, and Management

Previous chapters have discussed the packetized nature of ITV data, and chapter 10 amplifies this and describes the benefits of packets complying with the ATM specification. We have observed that these packets have destination address fields and data fields (along with other fields discussed in chapter 10) so that switching or routing of the packet can be accomplished automatically merely by looking at the packet address and sending it toward its destination(s). Routing is simple if the source of all program threads is centralized and shares one channel, and the threads are delivered to multiple destinations on that one channel as shown in Figure 9.1. The complexity in this figure is in only two places, the source of multiple program threads and the subscriber set top box. The source merely orchestrates the merging of packets and the decoders demultiplex (or unmerge) the desired signals.

The problem is compounded when networks are combined for purposes of making source material available to subscribers on one network, which is not available to subscribers on another network. For example, in Figure 9.2, four networks are shown interconnected. In this hypothetical scenario, it would be possible for each network to have program threads totally independent of each other, providing subscribers the ability to select 2800 different program threads. The worst-case scenario would be a totally homogenous combination of program destinations. This would create the greatest traffic problems, and it would create the most serious demands on the packet switch(es) and the delivery system. As the entropy of the source and destination combinations increases, so does the work the packet switch mechanism must perform, and so does the required number of media transport channels for the delivery system. As the number of channels increases, either the set top box must be able to select between multiple channels, or additional packet switches must be distributed near each cluster of subscribers to resolve the cabling dilemma. Therefore, Figure 9.2 is extremely simplified.

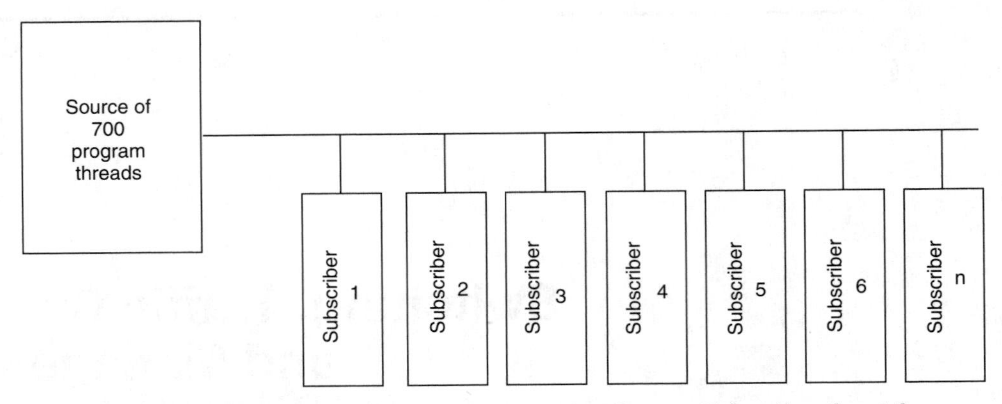

Figure 9.1 Packet switching application where all program threads are multiplexed together at the source and are individually demultiplexed by the subscriber set top box.

The packet switch must be able to route, in real time, all packets to all destinations. The packet switch must process packets in real time, not permitting any queues to build. As the network grows, the packet switch must be able to be distributed. The packet switch must operate in nonblocking mode so that any subscriber can access any movie any time. In this scenario, as the number of connected networks increases to create a desired number of video programs, the number of programs on each video server can decrease. For applications of fixed number of video programs, the number of interconnected networks can be inversely proportional to the size and complexity of the video servers at each of the hubs. Video servers are expensive, but so are ATM packet switches. This creates an environment for interesting price/performance trade-offs.

Issues other than cost and performance may come into play. When networks are interconnected in certain ways, the failure mechanisms likewise change. In a networked system, a major power failure, or facility fire, injured cabling or some other devastating event may permit most of the system to remain in operation with customers experiencing only some reduction in available programming capacity and only some customers experiencing total service outages. Numerous architectural variations need to be examined in terms of cost, reliability, ease of maintenance, expansibility, modularity, and future functions before selecting one approach.

Rings with connecting head ends and subsequent hubs make for interesting architecture. Due to bandwidth limitations, multiple overlapping rings will likely be required. Multiple rings require inter-ring switching, and this increases the switching system costs. This is depicted in Figure 9.3.

Some current network topology implementations have been discussed in chapter 4. The purpose of this brief chapter has been to shed light on the various economic analyses of possible system implementations; that is, network complexity versus video server capacity. The intent of this chapter was to illustrate the need for in-depth system analysis in making these decisions.

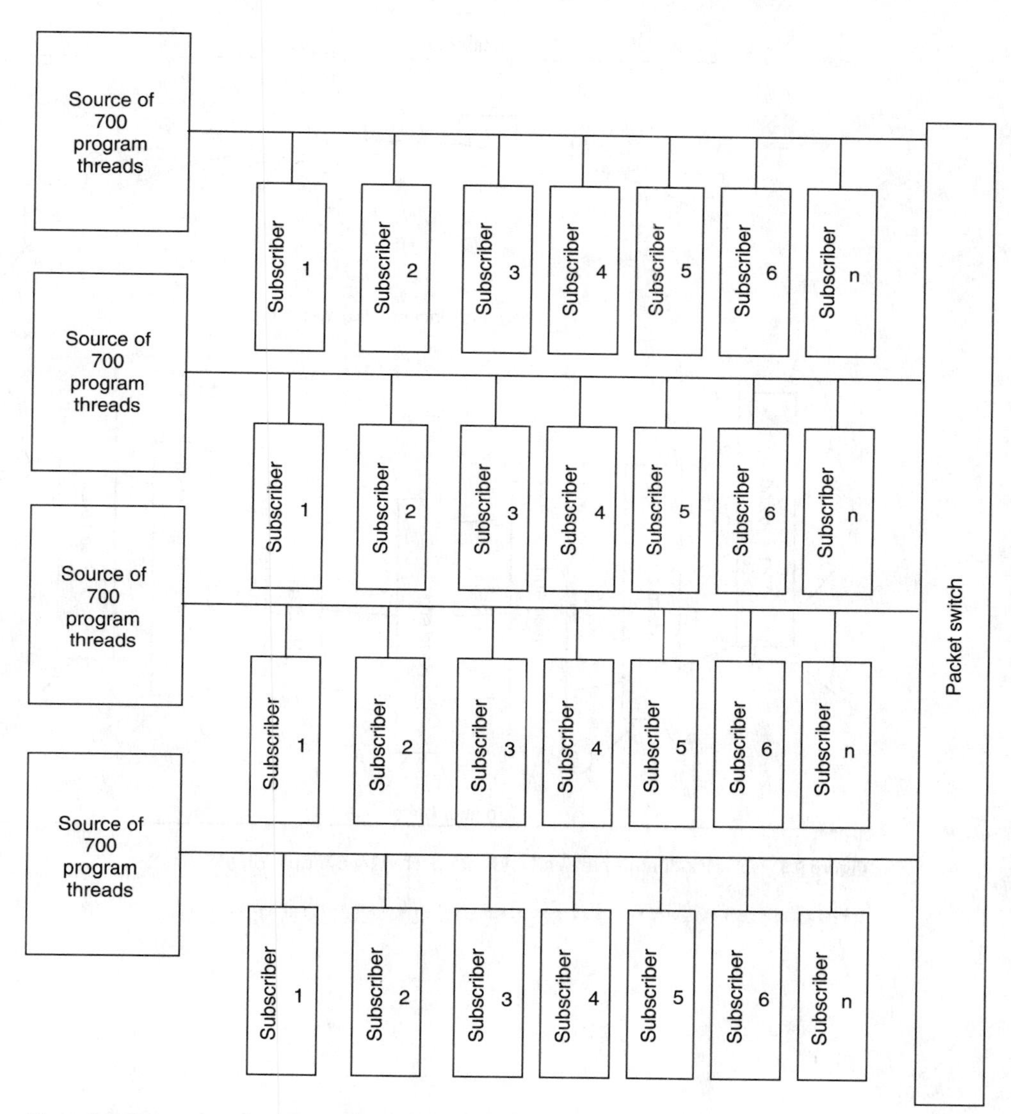

Figure 9.2 Four networks connected via a packet switch.

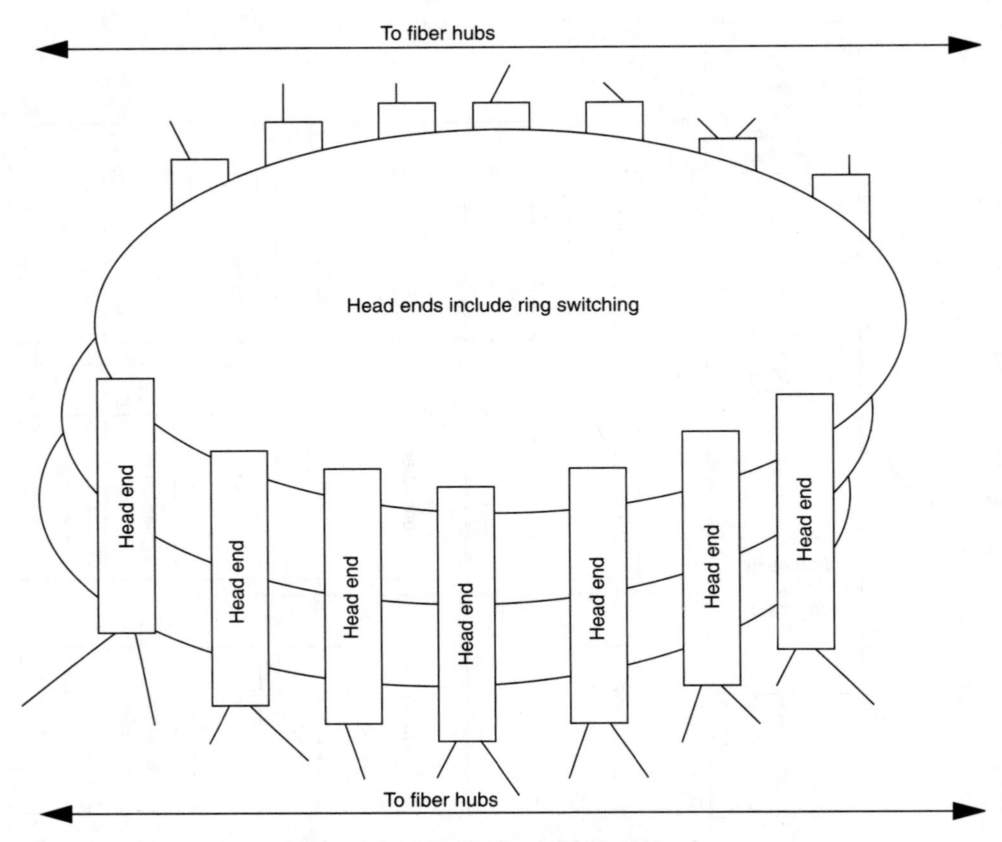

Figure 9.3 Network of multiple head ends connected by multiple rings.

Chapter

10

ITV Communications: Asynchronous Transfer Mode (ATM), Modulation, Enciphering, and Transmission

ATM is the first technology designed to explicitly merge compressed voice, data, and video communications into a common format. ATM provides sufficient performance to handle multimedia between people and machines. Compressed data requires the removal of redundant or nonessential information. A continuous flow of information will likely have varying degrees of redundancies or nonessential content. When this redundant or nonessential information has been removed, the previously synchronous bit stream becomes asynchronous. Packetizing variable-rate data streams is easily accommodated by ATM.

About ATM

ATM has the ability to scale from slow speeds to gigabyte-per-second transmission rates. Because of its packetized communications architecture, ATM can be readily compatible with LAN topologies such as FDDI and Ethernet, and WAN topologies such as ISDN or SONET (Synchronous Optical Data Network).

ATM provides an implicit multiplexer and demultiplexer function for asynchronous data packets. Packet multiplexing is performed by the transmitter. Packet demultiplexing is performed by the receiver. Routers add or subtract identifying headers or trailers to the data packets along the way. These headers and trailers are used for routing. Asynchronous transfer mode is a member of the general class of digital packet switching technologies that relay and route traffic by means of an address contained within the packet. Unlike the older packet technologies, such as X.25 or frame relay, ATM uses very short fixed-length packets referred to as cells.

ATM is differentiated with synchronous transfer mode (STM) in that STM describes the way the digital telephone network has operated since its inception under the guise of time division multiplexing (TDM). Most STM signals are the familiar DS1s and DS3s employed in today's digital telephone network.

ATM is an outgrowth of the TELCO's BISDN (Broadband Integrated Services Digital Network) standards. Telephone companies intended ATM to be carried on their synchronous optical network. The ATM cell shown in Figure 10.1 below is 53 bytes long. It consists of a 5-byte header containing addressing and control parameters and a fixed 48-byte information field. The ATM header represents a 9.4% overhead.

Packet technologies are attractive for data communications because they make better use of communications technologies than do TDM technologies. The common T-carrier systems and other TDM services are not routed by address, but by a time slot that is uniquely allocated when a call is dialed. TDM multiplexing has been used to good effect for assigning one user per time slot. But TDM technology assigns one time slot to each user whether or not it is required. Computer data requirements vary dramatically, depending on user requests for data. Assigning computers fixed time slots is wasteful and inefficient because of the dynamics of the environment.

However, packet switching protocols are quite efficient because when a large number of computers are networked together, traffic between the computers and their users is not likely to be consistent or correlated; that is, when one computer is not loading the network, another computer can use all of the network resources that are available. ATM is analogous to a group of people talking one at a time so as not to be garbled, but each individual being able to talk as long he or she wants. TDM has been a useful vehicle for voice transmission, because voice (which is not compressed in the time domain) will be sampled regularly, and these samples can be made to coincide with available synchronous time slots. The same is true of video. Sampled at more than the Nyquest rate, samples can be transmitted synchronously via a TDM system. When efficient temporal processing of voice and video occur to produce compression, a variable bit rate (or asynchronous transmission) will occur. It would be difficult to process on a synchronous network, but this is where ATM can excel.

For example, when compressed voice and/or video is packetized at an irregular rate and transmitted, it is necessary to time-stamp packets so that they can be reassembled in linear time. Voice and/or video traffic would suffer if the reconstituted data arrived in bunches and synchronization was lost. The result would be jerky voice or video. Packetized data also provides a statistical multiplexing function. When the bit rate of one channel increases, it is not necessary that the next channel's data rate will increase also, because these are uncorrelated channels. Because rapid scene changes using image

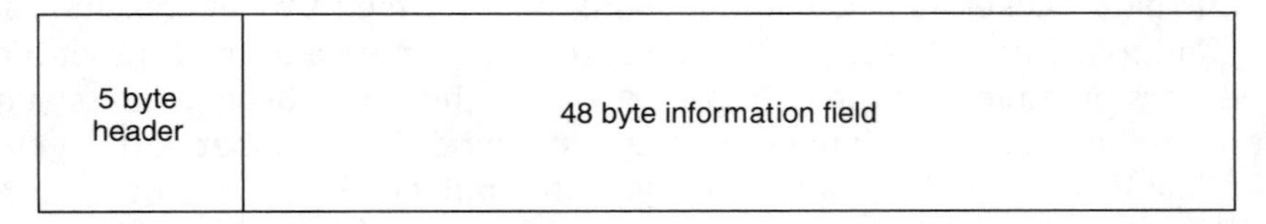

Figure 10.1 ATM cell structure.

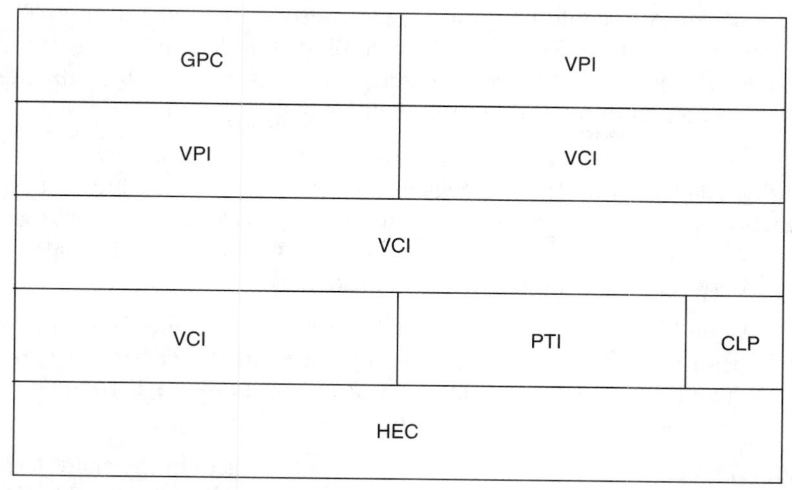

Figure 10.2 ATM header.

compression require more data bandwidth than slow-action pastoral scenes, more packets should be provided to the requiring channel during scene changes. Similarly, still or slowly moving images require fewer packets to produce the same quality video. Because there is statistical independence between different video sources (e.g., different movies), there is little likelihood of many channels requiring high bit rates at the same time, and the system bandwidth will not be exceeded.

The 53-byte length ATM cell is a compromise designed to make ATM viable for data, voice, video, and other real-time traffic that cannot tolerate randomly varying transmission delays. Computers can produce extremely long packets, but breaking long packets into small packets ensures that voice and video can be given priority and never wait more than one 53-byte cell before accessing the communications network. At 155.52 Mbps, a 53-byte cell occupies a time interval of 3.4 microseconds; therefore a 3-microsecond delay can be expected before the channel is accessed.

The header for ATM cells is formatted as shown in Figure 10.2.

For networking purposes, only the header is significant. It defines the cell's credibility and how the cell is to be delivered. The bytes are transmitted in an increasing order, starting with byte 1 of the header. Within the bytes, the bits are transmitted in a decreasing order starting with bit 8. The first bit transmitted is the most significant bit (MSB) in all the fields of the ATM cell. The ATM header is divided into the following fields:

GFC. The first four bits of the first byte contain the generic flow control field (GFC). It is used to control the flow of traffic across the user network interface (UNI) and into the network.

VPI/VCI. The next 24 bits make up the ATM cell address. This 3-byte field (four bits displaced) is divided into two subfields. The first byte contains the virtual path identifier (VPI), and the second two bytes make up the virtual channel identifier (VCI). In an interactive CATV system, the VCI could be the TV set top box real address, while the VPI could be the program address. A value 7 of the PTI field (currently reserved for future use) could be employed to discriminate between VPI and VCI. Thus, a 24-bit real addressing scheme can be used for set top box identification (over 16 million different address possibilities in one system).

PT. The 3-bit payload type (PT) field is used to distinguish between user data and network management cells. ATM cells will be used to transport different types of user information that may require different handling by network or terminating equipment. Cells also will be used to transfer operations and maintenance information across the network between users and the service provider.

CLP. A one-bit cell-loss priority (CLP) field is used to mark the cell's priority and is set by the user. The bit indicates the eligibility of the cell for discarding by the network under conditions of congestion. If the bit is set to 1, the cell may be discarded by the network should traffic conditions dictate.

HEC. The header error-check (HEC) byte (byte 5) uses cyclic codes to perform error correction across the first four bytes of the header. It is designed to detect multiple header errors and correct single-bit errors, and it also provides the cell delineation function. The main purpose of this field is to provide protection against misdelivery of cells due to addressing errors. The HEC byte does not validate the content of the 48-byte information field.

Information. The information field is the 48-byte field that carries either discrete and/or composite audio, video, and control data.

The system requirements for NVOD or TVOD can impact communications addressing and storage (see chapter 7). TVOD requires a separate communications path to each user. NVOD permits program sharing and subsequent virtual address sharing. NVOD is a broadcast mechanism permitting multiple subscribers to share one virtual address. In other words, in an NVOD system, 500 people could, for example, share one program (one virtual address) being broadcast to them. In a TVOD system, a program would be sent as 500 different sets of cells to the 500 subscribers via 500 different addresses. The TVOD system in this case requires 500 times more bandwidth and is more complicated and costly to transmit, transport, and decode. An NVOD system has one extra step of complexity; namely, after subscriber program selection, a virtual address must be sent to each real-addressed box requesting the particular program. This is an inexpensive function.

Large NVOD systems will broadcast probably hundreds of programs or program threads while large TVOD systems will likely transmit thousands. Hundreds of programs can be handled by existing CATV coaxial cable while thousands of programs will require fiber-optic cable and distributed ATM to coaxial switching. It will cost significantly more to transmit TVOD than NVOD. However, a system could be designed to permit concurrent combinations of the two.

Now that we have looked at what appears to be the signaling protocol, we are ready to address the following problems:

- What is a possible band plan? How should the system spectrum be used?
- How might a 1-GHz, 275-channel hybrid analog/digital CATV system appear?

- If ATM signaling is to be placed on the CATV cable, how should it be modulated?
- How many program or program threads should be combined, and what bandwidth should the ATM signal occupy? How does this affect video quality?
- Should the ATM signaling be compliant with TELCO specifications?
- What effect does standards compliance have on system complexity, reliability, maintainability, cost, and future functionality?
- How is the bandwidth used on existing CATV systems?
- What level of security should be provided?
- What new features will be added to the CATV system of the future (e.g., telephone, shopping, video-phone, games, education, residential monitoring and control . . .)?
- What are the business applications?

Band Plan and Spectrum Utilization for Hybrid 275-Channel System

Several candidate frequency spectrum utilization plans have been conceived. The following plan (Figure 10.3) represents the allocation of 75 conventional analog television channels in the VHF range between 50 MHz and 550 MHz, and 200 to 400 compressed digital channels between 550 MHz and 1 GHz. This plan shows the HF band (5 MHz to 50 MHz) being used as the return path to provide interactivity. These digital channels are to be used for pay-per-view (PPV) TV, interactive TV, HDTV, VOD, home shopping, etc.

Figure 10.4 represents a CATV system broadcasting 75 analog channels on 50 MHz through 440 MHz with 200 (could be more) digital interactive television channels in

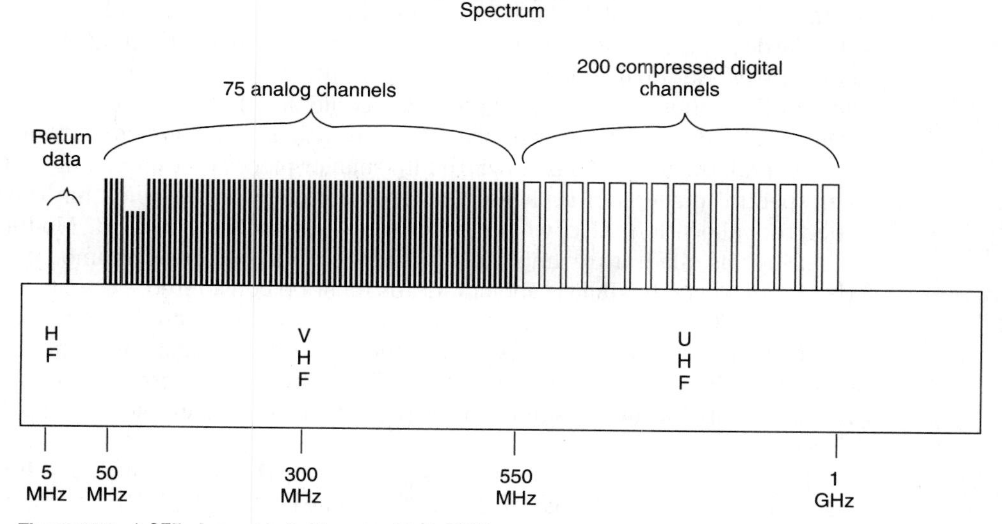

Figure 10.3 A 275-channel hybrid analog/digital TV system spectrum plan (*Time-Warner Corporation*).

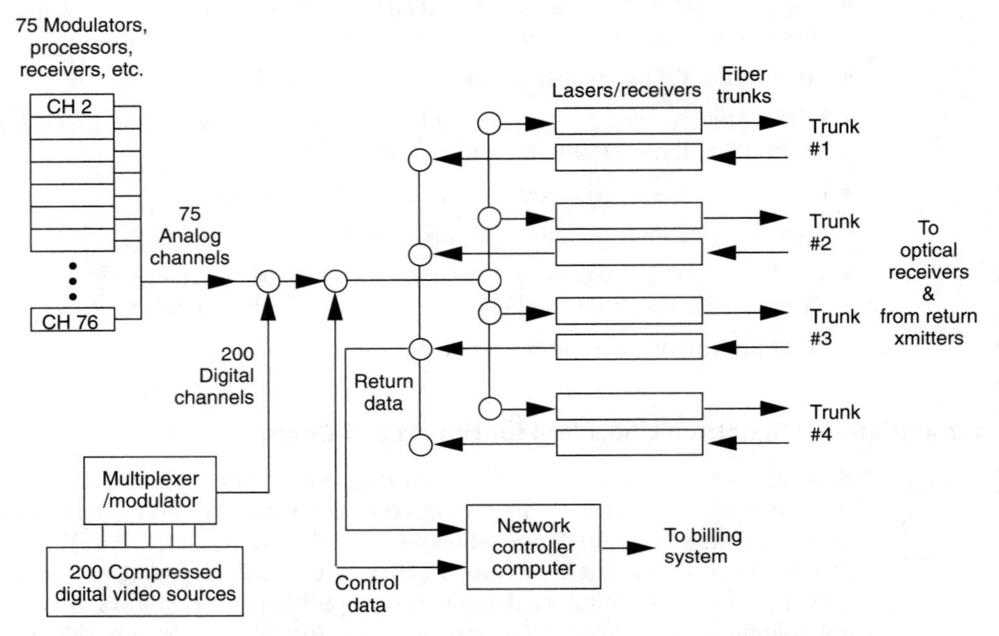

Figure 10.4 A CATV system functional block diagram for the 275-channel hybrid system using the spectrum described in Figure 10.3. (*Time-Warner Corporation*).

the frequency range from 440 MHz to 1 GHz. The question should be raised as to what digital plan should be employed to place those digital channels in that allocated frequency spectrum.

Imposing the digital TV audio and video in an ATM format would be useful from a standards point of view and would maintain compatibility with numerous systems including the National Information Infrastructure (NII). The next question is: How should the bandwidth be employed? One could simply take 200 channels of digital television and place it in the allocated bandwidth as one composite element, or one could break the allocated bandwidth into smaller pieces of, for example, 10 channels per spectral piece, and replicate that spectral piece 20 times within the 200-MHz spectrum allocation. Also, what modulation system should be used? Modulation systems such as quadrature amplitude modulation (QAM) can accommodate 4 bits per Hz. Again, costs, flexibility, and standards must be considered.

Some CATV system operators preparing for digital communications are considering digital systems placing 4 digital channels where previously only one existed, i.e., 4 channels in a 5-MHz spectral interval. The chart below depicts video quality as a function of its bit rate, assuming 4-bit-per-Hz encoding is employed. This is a realistic expectation.

The new Pacific Bell network will operate over the 50-MHz-to-750-MHz range in the forward direction (central office to the home) and in the 5-MHz-to-40-MHz range in the reverse direction, leaving a guard band between 40 MHz and 50 MHz. They will have multiple local switching nodes to provide for the electronic transport to a local loop of over 250 programs. These nodes connect to fiber-optic cable on one side, and to coaxial cable serving up to 500 homes on the other side.

It is obvious from Table 10.1 that 4 bits per Hz can provide anywhere from 2 to 10 digital channels, depending on the quality of video transmitted.

We now can examine the possibilities when a rate of 6 bits per hertz is employed, as explained in Table 10.2 below. This illustrates that 3.5 to 16 digital television channels can be made available, depending on desired quality. One could mix different quality channels to develop many different mixes in a 6-MHz spectrum space.

Telephone companies are subscribing to ATM; the NII will probably convey ATM formatted data, and CATV companies are examining ATM as a formatting mechanism, all probably because of the usefulness, availability, and standards compliance of the technology. It is then reasonable to examine porting mechanisms such as T-Carrier, so that if necessary the signaling could be accommodated across the major wired networks.

Figure 10.5 represents the characteristics of three transmission methodologies. It can be seen that a 6-MHz CATV channel could accommodate 5 broadcast-quality digital television channels. Similarly, OC-3 could accommodate 32 channels with video quality in the range of VHS and broadcast. A C-band satellite transponder can provide 6 broadcast quality programs.

Digital circuits for phase locking, decoding, and correcting errors become expensive above 275 Mbps, because very-high-speed logic circuits are required. Decoding and processing is much less expensive under 125 Mbps. Therefore, there exists a rationale to place groups of data modulated in various parts of the available carrier's RF spectrum where they can subsequently be demodulated and processed as independent manageable frequency intervals.

TABLE 10.1 Expectation of 6-MHz Channel Utilization Using 4-Bit-per-Hertz Encoding

MPEG Quality	Data rate	# programs/ 6-MHz channel	# programs/ 500 MHz
VHS	1.5 Mbps	10	880
Broadcast	5.0 Mbps	3	274
Studio	7.0 Mbps	2	192

TABLE 10.2 Programs Possible with 6-Bit-per-Hertz Encoding

MPEG Quality	Data rate	# programs/ 6-MHz channel	# programs/ 500 MHz
VHS	1.5 Mbps	16	1333
Broadcast	5.0 Mbps	5	416
Studio	7.0 Mbps	3.5	291

Channel Medium	# programs	program bandwidth Mb/s	required bandwidth Mb/s	transport bandwidth Mb/s	% bandwith utilization	% overhead
6-MHz CATV channel	5	5.2	26	27.10	95.94%	4.06%
OC-3 fiber channel	32	3.9	124.8	128.20	97.35%	2.65%
C-band satellite transponder	6	5.6	33.5	34.70	96.54%	3.46%

Figure 10.5 Bandwidth utilization for 3 predominant transmission methodologies. This information was derived from the MPEG specification referred to as "ISO/IEC # 13818," published 6/10/94.

Figure 10.6 shows a composite video server, ATM signal multiplexer, and modulation system capable of accommodating any of the channel media shown in Figure 10.5.

Figure 10.6 is significant because it takes advantage of disk formatting and encoding and uses the byte serial data in an optimum way all the way to the ATM spigot and to the network. The system is capable of producing hundreds of transport streams.

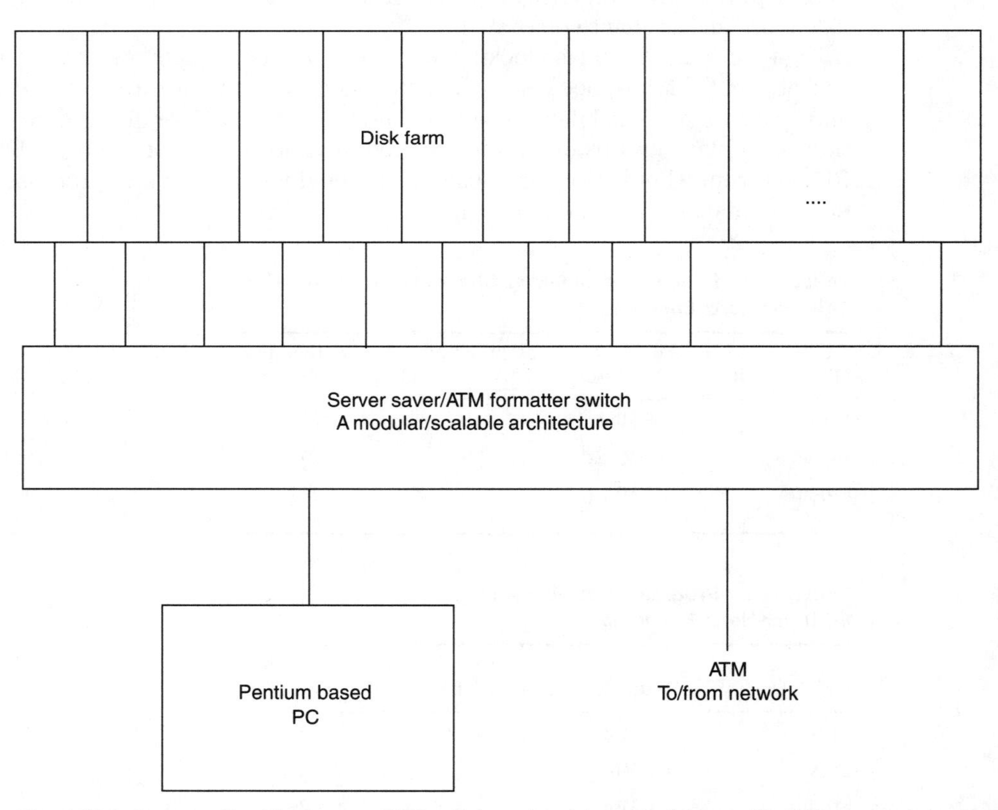

Figure 10.6 A composite video server, ATM switch, and spectrum modulator with ability to produce signaling mediums as represented in Figure 10.5.

Conclusion

ATM's scalability provides the seamless integration from WAN to LAN without any requirement for protocol conversion at the user-to-network interface. In fact, ATM is the first technology to gain broad support of desktop, computer, internetworking hub, and telecom transmission/switch developers. Manufacturers of public network phone equipment and CATV companies have widely accepted ATM switching and interface technologies, and they are developing a large-scale public-network ATM infrastructure.

Several factors are increasing ATM's use in public networks. Because the mechanisms for SONET/SDH transport are stable and well-understood, public network developers can implement ATM technology over SONET/SDH frame structures with confidence. In addition, ATM's technical and marketing advantages are driving increased competition among local and long-distance service providers. This increased competition means that public operating companies are investing in ATM equipment at an unusually rapid pace. Cooperative development among CATV and telecom OEMs and semiconductor manufacturers is resulting in the production of ATM switches that provide aggressive price, performance, and power ratios.

The global impact of this technology literally demands a commitment of resources and expertise that exceeds the capabilities of even the largest multinational, high-technology firms. Only through the cooperative development efforts of technology innovators can ATM come to the market at a price that will ensure rapid assimilation of this protocol. Cost-effectiveness is a key factor for the successful adoption of this standard, and a prime example of this trend toward innovation through cooperation is gaining momentum.

As a local area network transmission format, ATM offers greater bandwidth than the current assortment of shared-bandwidth network technologies. The most common shared protocols include Ethernet, Token Ring, Fiber Distributed Data Interface (FDDI), and 100BaseT, which is a higher-bandwidth version of Ethernet. When used with SONET framing structures, ATM can provide dedicated and scalable bandwidth from 24 Mbps to 2.4 Gbps and beyond. For desktop LAN applications, ATM can provide 155 Mbps through wire or fiber-optic cable, while in contrast Ethernet transfers only 10 Mbps and Token Ring transfers 16 Mbps, while 100BaseT and FDDI each deliver 100 Mbps performance.

With the advent of the ATM Forum, ATM is the first technology to experience the level of consensus that comes from a symposium of more than 400 companies working together to support the computer, internetworking, telecommunications, software, hardware, semiconductor, and commercial industries. Although the ATM Forum is not itself a standards body, it facilitates the early development of ATM standards by working closely with organizations such as ANSI, IEEE, and ITU-TSS.

The ATM Forum has initially focused on ATM's user-to-network interface to define four physical-connection protocols. Physical-layer semiconductors are responsible for all cell manipulation between the physical media (fiber or twisted-pair cable) and the ATM network's segmentation/reassembly adaptation boundaries.

ATM simply outperforms the older networking communication technologies. Ethernet and Token Ring networks were designed to transfer low-rate but bursty data between terminals, servers, printers, and other peripherals. As a result, neither can handle the constant bit-rate traffic generated by voice and video applications. And although FDDI and 100BaseT specifications claim speeds reaching 100 Mbps, actual transmission speeds per user are usually lower, because network users must share the network's bandwidth. Furthermore, FDDI has been plagued by expensive port costs and a lack of finalized standards. ATM's relatively short, fixed-length cells lower the cost of data processing as compared to FDDI.

It can be deduced from this chapter that ATM has the very high probability of becoming the standard bearer for CATV companies, TELCOs, LANs, and WANs because of its high performance and reliability and low cost. The preferred communications mechanism for ITV, PPV, VOD, vision phones, and multimedia is ATM.

Chapter

11

ITV Set Top Box Requirements and Architecture

The TV set top box is the device that controls the video server in modern interactive TV systems. The set top box has changed from a multifrequency tuner and descrambler into a controlling front end for a massive on-line database of movies, multimedia events, news, etc. Companies such as AT&T Network Systems, IBM, Hewlett-Packard, and Apple are attempting to usurp the business of traditional set top box manufacturers, which include General Instruments, Scientific Atlanta, and Zenith, because of their self-perceived capabilities in the data distribution and control business.

US House Bill 3636, authored by Representative Ed Markey, calls for the FCC to commence a proceeding and report to Congress on the importance of open and accessible systems, and the security concerns between the network interface and the set top box. Outside Capitol Hill, the open set top special interest group of the Video Electronics Standards Association (VESA) is proposing a standard architecture. The VESA committee is dominated by PC makers and semiconductor companies.

An open-architecture ITV set top box can be a simple low-cost box containing an inexpensive microprocessor controlling specialized VLSI chips providing discrete functions. These include: a digital tuner and an ATM processing unit or an ADSL interface, image and acoustic decompression, NTSC, PAL, SECAM, or RGB encoding, channel 3 or 4 modulation, bidirectional signaling, interface to a remote control, and memory, along with an expansion interface for optional and future features. The advantage of such an open architecture is that the box does not need to know what kind of service the subscriber requires, or what optional features, peripherals, or computers are connected. The simple box would only have to query, respond, route, decode, and decompress data, and handle transactions and restrictions.

The ramifications of such an open architecture are massive. It would reduce the television receiver to a simple monitor, a concept known as convergence, but a concept adverse to the Electronic Industries Association (EIA). The concept of modularity, in the case of set top boxes, will actually reduce the costs of (and permit higher quality) monitors suitable for the multiple purposes of standard TV viewing, NTSC/PAL/SECAM-independent TV, HDTV, and computer use.

ATM is an effective means of communicating to the home, to the business, and within the business that employs LAN, WAN, or MAN topologies. It consolidates digital communications technologies and will bring forth new economies of immense scale.

Open architecture provides the consumer the cost-effective option of owning his own set top box equipment without fear of future incompatibility or rapid obsolescence. It permits the insertion of the open architecture set top box functions inside new television designs. Such new televisions would be NTSC/PAL/SECAM-independent. The open-architecture design methodology provides the most utility to the consumer because:

- It does not freeze the technology.
- It efficiently connects to the network.
- It offers the most flexibility and functionality.

Electronic Program Guides (EPG) functionality will be included in the set top boxes, providing on-screen graphics and the capability for real-time interactivity and menu navigation. EPG vendors include StarSight Telecast, TV Guide On Screen, and Preview Guide.

The set top box of the future should accommodate upstream data rates from 1.5 MHz (VHS quality) to 10 MHz (studio quality). The CATV set top units should have digitally controlled tuners permitting a variety of ATM transport mechanisms, from OC3 (155 Mbps on a single 40-MHz wide channel) to a 24 Mbps on a 5-MHz channel. The out-of-band data signaling should be able to be placed either in the low band (5 MHz to 45 MHz) or on a high band above 500 MHz. To be economically viable, the set top box should be saleable for under $150.

A design for a fully featured CATV 150 set top box is shown in Figure 11.1.

The set top box is *not* merely a tuner, as were previous set top boxes. Rather, it is a remote control unit for the video server, to which it connects through an ATM network. This is the distinction between this unit and pre-ITV units. It provides bilateral, full duplex communications to the video server.

The ensuing Figure 11.2 depicts the almost identical set top box for telephone company interface to standard 22-gauge POTS twisted pair. The similarity should be readily apparent. It is possible to manufacture one PC board, populated only slightly differently for TELCO or CATV applications. Perhaps the only difference could be a plug-compatible TELCO/CATV VLSI front-end device.

The signaling plan for TELCO and CATV applications is different. In the TELCO application, a separate pair of twisted wire pairs runs from the central office (CO) to each different subscriber, i.e., each subscriber has his or her own dedicated pair of wires. Somewhere between 1.5 and 6 Mbps can be transmitted over each separate pair of TELCO twisted pairs. This bandwidth is adequate for one compressed signal capable of producing video ranging from VHS- to broadcast-quality TV reception.

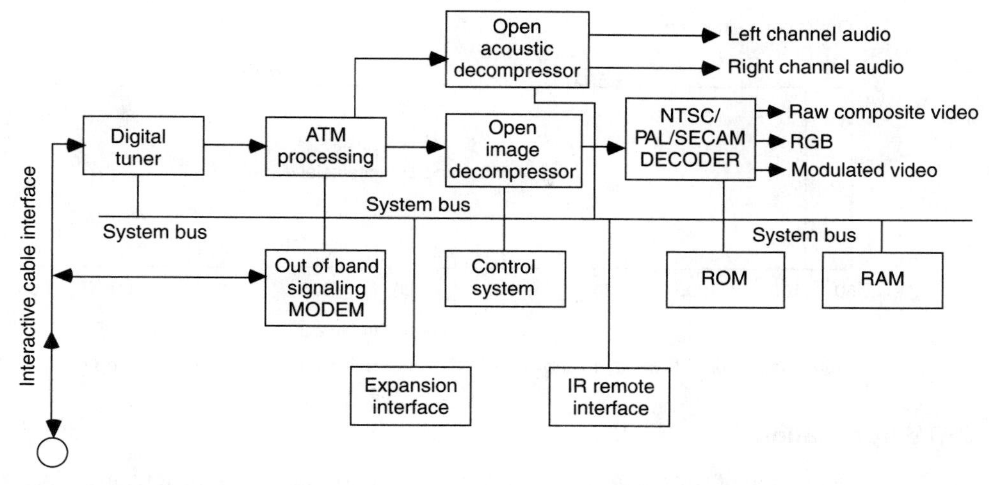

Figure 11.1 Functional block diagram for a typical CATV set top box.

Between 250 and 500 TELCO twisted pairs would be required to equal the bandwidth of one CATV coaxial cable. A single-cable CATV system can deliver from 500 to 1000 composite program threads to a community, but the TELCO could deliver more! The CATV broadcasts all threads over a single cable. The TELCO delivers only one program over each twisted pair.

We will examine how bandwidth is employed in these two variant systems. First, CATV bandwidth allocation is shown in Figure 11.3, and TELCO ADSL bandwidth is represented in Figure 11.4.

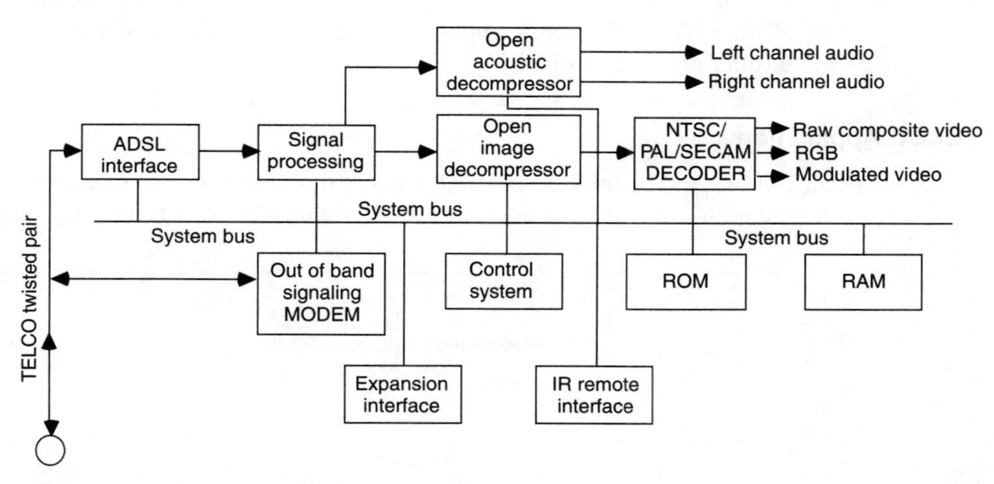

Figure 11.2 Functional block diagram for a typical TELCO set top box.

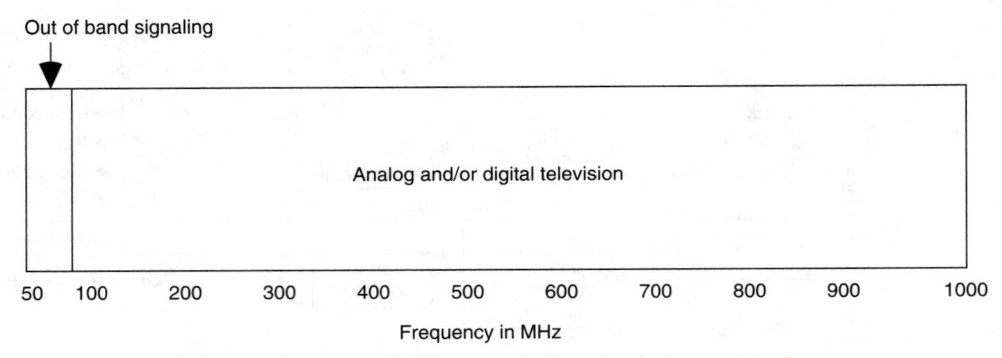

Figure 11.3 Possible scenario showing CATV communications band plan to set top box.

CATV Application

A portion of the cable spectrum will be used for out-of-band signaling. This is for use by the set top box to communicate with the video server and will contain such information as the viewers' menu selections, hardware/line diagnostic status, requests for program downloads, etc. At the present time, various candidate frequency allocations are being considered including the unused low band (from 10 to 40 MHz). However, out-of-band allocations are also being considered in the high end of the spectrum (see chapter 7).

The communications between the set top box and the video server are highly asymmetric; that is, information from the head end to each set top box requires many times more capacity than does the return path, depending on the level of interactivity. Similarly, out-of-band signaling to the set top box has similarly small bandwidth requirements. This is why only a small portion of the available bandwidth is used for out-of-band signaling, typically 5 to 10 percent. By employing digitally programmable tuners and out-of-band signaling modems, the cable characteristics can be either automatically determined by the set top box, or defined by a cable operator so that one set top box can operate with various configurations of cable systems. The STB for the CATV application (unlike the ADSL application) requires a digital tuner and ATM front end.

TELCO Application

The TELCO ADSL application was designed to permit concurrent voice, compressed audio, compressed video, and return data to co-exist. It is called Asymmetric Data

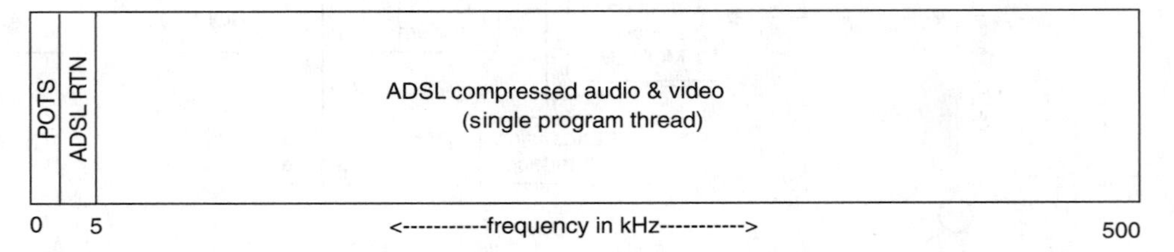

Figure 11.4 Possible scenario showing TELCO communications band plan to set top box.

Subscriber Loop (ADSL) because of the highly asymmetric nature of data transmission. Plain Old Telephone Service (POTS) occupies the first approximate 5 kHz of bandwidth. The return path occupies the second 10 kHz of bandwidth, with that balance being left for compressed audio and video.

The TELCO set top box requires no digital tuner because only one compressed audio-video signal is transmitted to it, unlike the CATV set top box, which must select from hundreds of compressed audio-video signals carried on the CATV coaxial cable. This implies that in a household with many TVs, in order for several viewers to watch different programs simultaneously, a pair of twisted pair cables might be required for each TV program. The functional electronic building blocks comprising the set top box are subsequently described.

Control system (microprocessor)

The control system, or system microprocessor, is the heart of the set top box. It receives its start-up functions (bootstrap) from the attached Read Only Memory (ROM) and stores dynamic programs and data in its Random Access Memory (RAM). The system bus that connects the functional building blocks permits the control system to orchestrate the other building blocks to produce the total system functions. These functional electronic building blocks are logical candidates for VLSI implementation with the possibility that more than one of these functions can be combined into one VLSI device.

Digital tuner

The Digital Tuner (not part of the TELCO application) receives instructions from the control system advising it about the type of transport layer, the bandwidth, the frequency, the demodulation scheme, and the decoding procedure. When initialized, it produces a digital data stream for the ATM processing function. This digital data stream may include multiple programs or program threads. For each frequency the digital tuner selects, different groups of concurrent programs are selected. Each group may vary from as few as 2 programs to as many as 100 programs per frequency interval.

ATM processor

The ATM processing function (also not part of the TELCO application) receives packets of data from the digital tuner. The ATM processor receives information from the control system directing it to select certain data packets. For example, the control system may direct the ATM processor to select only data packets with certain virtual or real addresses, while discarding all other packets. Specific real or virtual packet addresses will correlate to specific programs.

The ATM processor separates the selected data into three streams: one video stream, one audio stream, and one data stream available to the control system. The audio stream is fed to the acoustic decompressor while the video stream is sent to the image decompressor. Control information from the head end defines decoding algorithms, viewing restrictions, image resolutions, aspect ratios, and so on. The

ATM processor also collects status regarding errors in transmission and submits this information to the control system when requested, so that the control system can communicate this information to the head end.

ADSL interface

Part of the TELCO interface only, the ADSL interface retrieves data from and places data on the TELCO twisted pairs. It makes the compressed digital audio and video signals available to the appropriate decompression circuits.

Open image decompressor

The open image compressor receives information from the control system advising it on how to decompress the image. It then receives compressed video data from the ATM processor and decompresses it for transmission to the NTSC/PAL/SECAM decoder.

The open image decompressor receives special image decompression firmware from the control system, enabling it to execute a wide variety of image decompression algorithms. The open image decompressor function is a specially designed programmable digital image processing VLSI device that, once loaded with appropriate firmware-based algorithms, uses that firmware to decompress the previously compressed video. These firmware decompression algorithms can be updated on a program-by-program basis (or even a scene-to-scene basis), permitting the application of the best decompression technology for the scene being displayed.

Open acoustic decompressor

The acoustic processor function receives decompression algorithms from the control system and compressed digital acoustic data from the ATM processor. It decompresses the acoustic data and presents it to the TV set as modulated or unmodulated stereo signals.

NTSC/PAL/SECAM encoder

The NTSC/PAL/SECAM encoder creates video signals as required for the TV set or monitor. It provides compatibility for the basic worldwide standard video signal interfaces.

Out-of-band signaling modem

The out-of-band signaling modem provides the interface between the set top box's control system and the video server at the CATV head end. This function is common to the TELCO application. It provides full duplex digital data communications to the set top box. In the CATV application, it defines which real or virtual addresses should be decoded. In both applications, it provides instructions on how to decompress, descramble, and decode them, what the program ratings are, the timing of the program threads, menuing and navigation information, bandwidth planning, Electronic Programming Guide (EPG) information, pricing, bootstrapping, etc. This subsystem communicates customer requests, system status, diagnostic responses, etc. to the head end.

IR remote interface

The IR remote interface function provides an interface to popular IR remote control devices, including the one-button point-and-click devices. The IR interface includes the full duplex capabilities required for the one-button point-and-click remotes.

Expansion interface

The expansion interface provides the ability for the control system to extend its bus so that it may communicate with other devices such as game ports, SCSI controllers, peripherals, and other computers. In this way, the basic cost of the set top box is kept as low as possible.

Telephone interface

A conventional RJ-11 connector allows plugging a standard telephone into the set top box. In the TELCO application, it will connect the phone to the house wiring. In the CATV application, it allows the user to engage in *TELCO-style* voice communications with anyone via the CATV network, using a very small percentage of the available CATV bandwidth. Of course, this assumes the appropriate ATM switching facilities at the CATV head end to permit them to communicate. This will allow for additional revenue sources for CATV operators and will undoubtedly be a source of some concern on the part of the TELCOs.

A common set top box design for both TELCO and CATV applications permits additional economies of scale without sacrificing cost or function. All the above functions are orchestrated locally by the set top box control system. However, because this control system receives its programs by bootstrapping from the head end, the set top box is very much under the control of the CATV system operator. Because of the open architecture, the set top box can operate with a variety of operating protocols. The set top box configured above assumes that the functional electronic building blocks attach to the control system bus interface on a programmable-address basis, each device responding to a block of device addresses.

System Operation

The video server and ATM switching and transport mechanisms at the head end create and compress data, packetize it, and send it down the CATV system. The ATM protocol provides address space for up to 16 million distinct real or virtual addresses. The set top box can decode this entire address space. Each set top box has its own preprogrammed, fixed real address stored in its own unique ROM. When a set top box requests a program, it attaches its real address to its request. The video server system responds with a virtual address that the set top box uses for the duration of the program.

Because several set top boxes can decode the same virtual addresses at the same time, one program or program thread can be broadcast to many set top boxes. While the box is idle or not receiving any program material, it exclusively uses its real address for communications. When it is in a program reception mode, it can respond to both real and virtual addresses.

A unique virtual address is attached to a specific time sequence of each program. All viewers viewing the same program with the same start time are viewing the same data packets. Employing this technique and standard CATV coaxial cable, 500 to 1000 program threads could be transmitted to (and decoded by) many thousands of set top boxes.

The set top box should permit the selective reception of programs with resolutions varying from VHS quality (~ 1.5 Mbps) to HDTV (~ 12 Mbps), and it should be able to operate with fixed bit-rate or variable bit-rate video. It should respond to all status inquiries, including those regarding transmission errors.

Set top box (STB) bootstrapping

The open architecture of the STB permits compatibility to a wide variety of CATV systems with supporting different ATM transport layers and image decompression and processing, but it requires either a custom ROM for each variant the STB supports, or a bootstrap in ROM that permits downloading the system functionality into RAM for execution. The bootable STB system therefore requires, as a minimum, a small custom ROM with instructions on how to perform the primitive communications to request the loading of STB firmware into RAM. Such bootstrapping generally requires knowledge of the STB model identification and serial number. A simple query between STB and head end is shown as Figure 11.5. Figures 11.6 through 11.11 show flow diagrams for other program processes carried out by the set top box.

The STB should be capable of presenting sequences of navigable menus of current or future programming or media events. These menus and schedules should be created for distribution by the head end.

The STB may request selection to the head end, or the head end may initiate a connection to the STB. Connection requests by the head end to an STB may be specific (real) or general (virtual), but a virtual addressing sequence always follows a real addressing sequence. Real addressing by the head end can be for purposes of collecting or validating STB requests, status, usage data, or diagnostic information, or for sending a virtual address to the STB.

STB controls

As previously mentioned, the system may have either of two different types of remote controls: either the conventional, inexpensive multiple-key IR remote, such as are in common use today for TV, VCR, and audio systems; or the more user-friendly one-button remote control previously described in chapter 2.

The video server has information concerning the configuration of each set top box, uniquely identified by its real address. It downloads remote control interface software for the appropriate type of remote during the initialization process described above. In the case of the conventional remote, the user may request a menu by pressing a particular two-key combination. For the one-button remote, the act of picking up the control activates a sensor that automatically sends a request to the STB to display a menu.

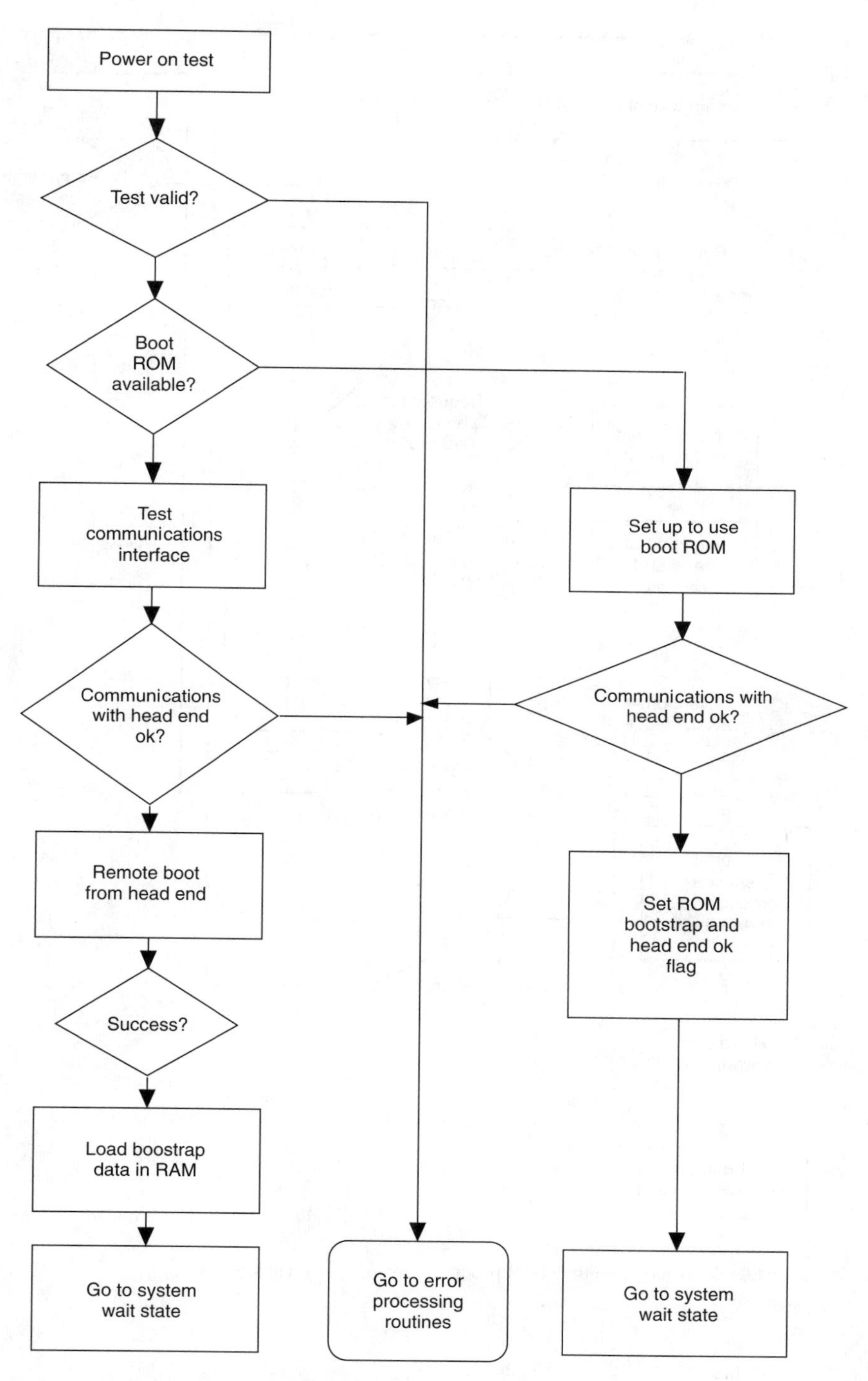

Figure 11.5 Power on/bootstrapping: STB polling.

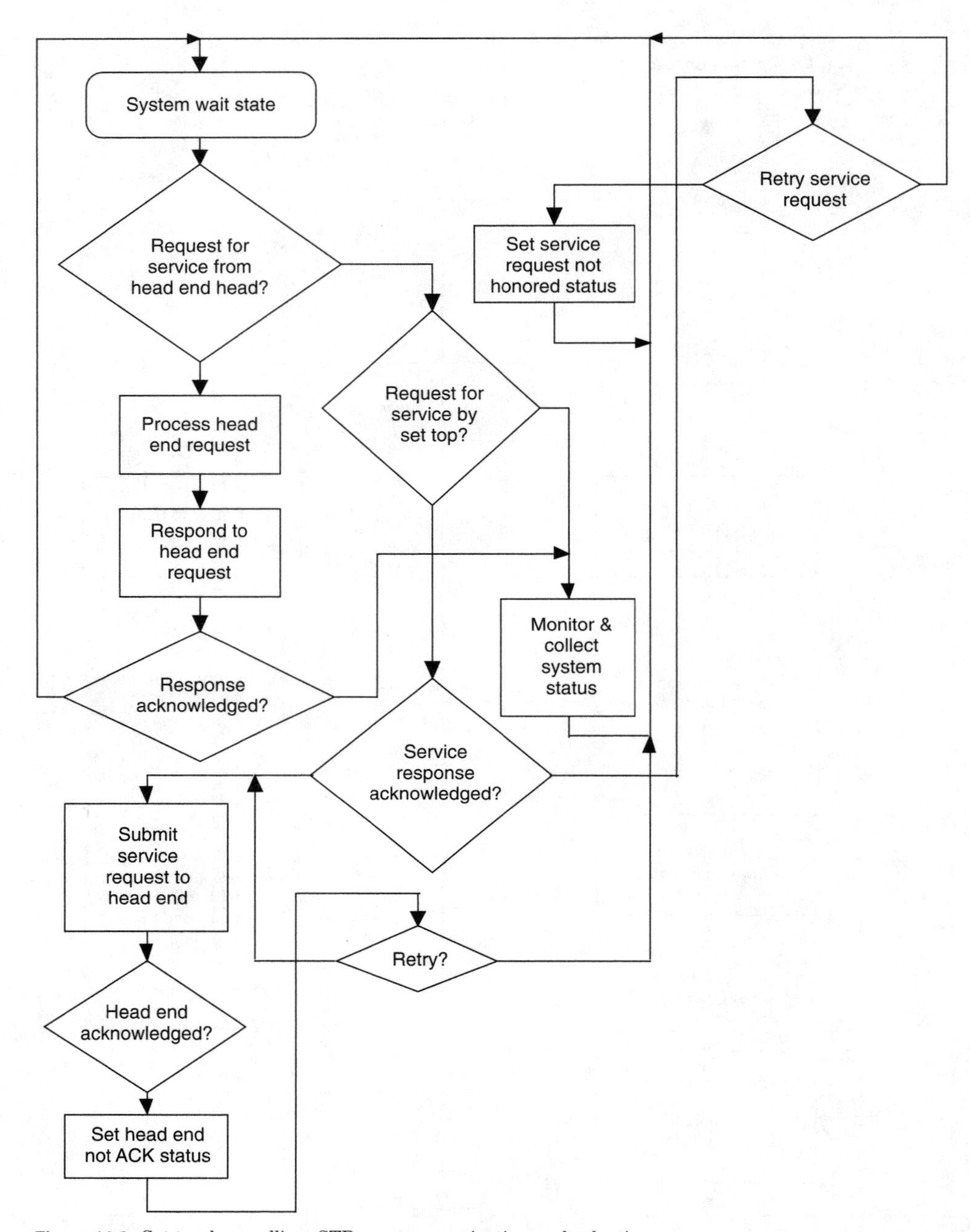

Figure 11.6 Set top box polling: STB program navigation and selection.

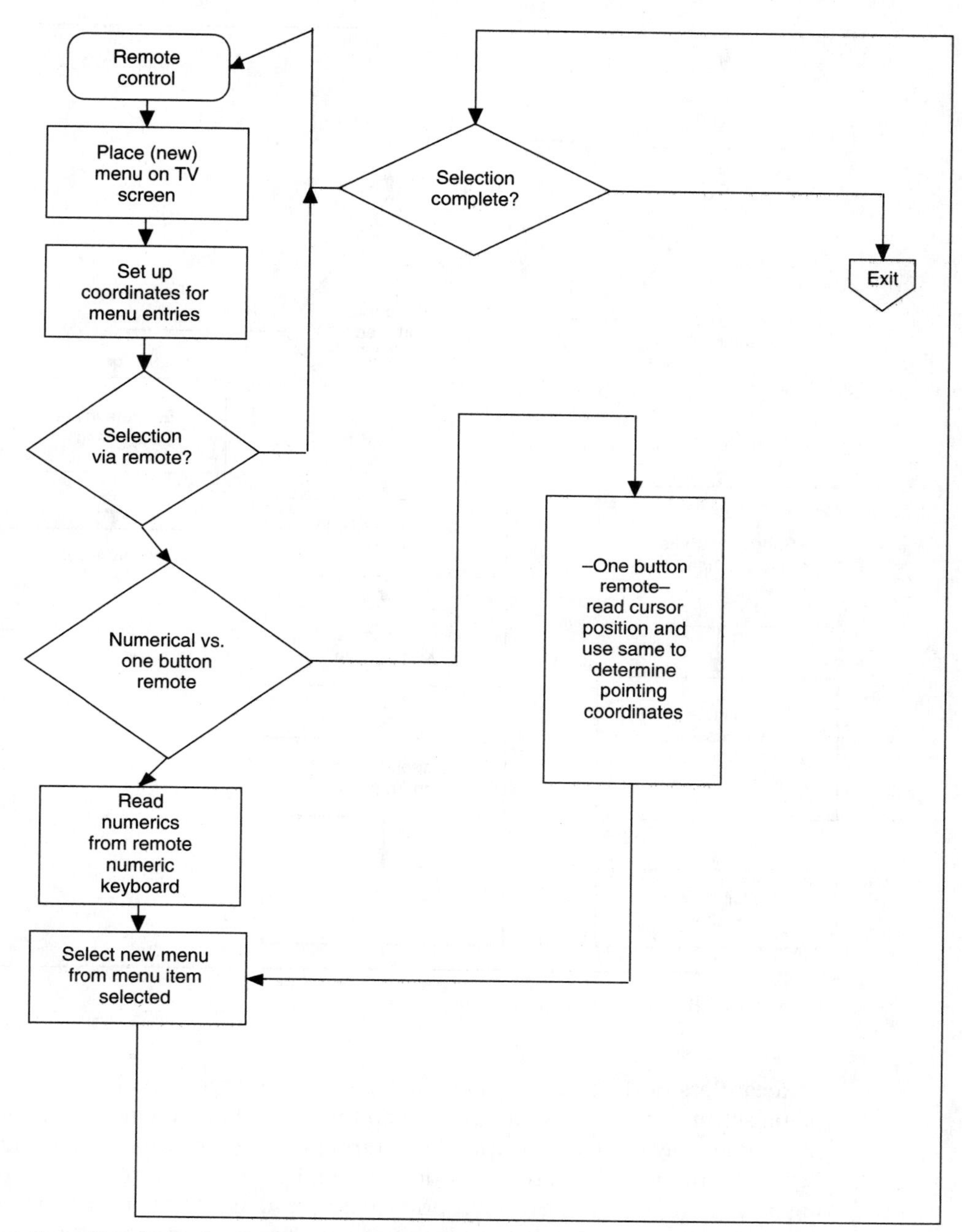

Figure 11.7 Set top box program for navigation and selection.

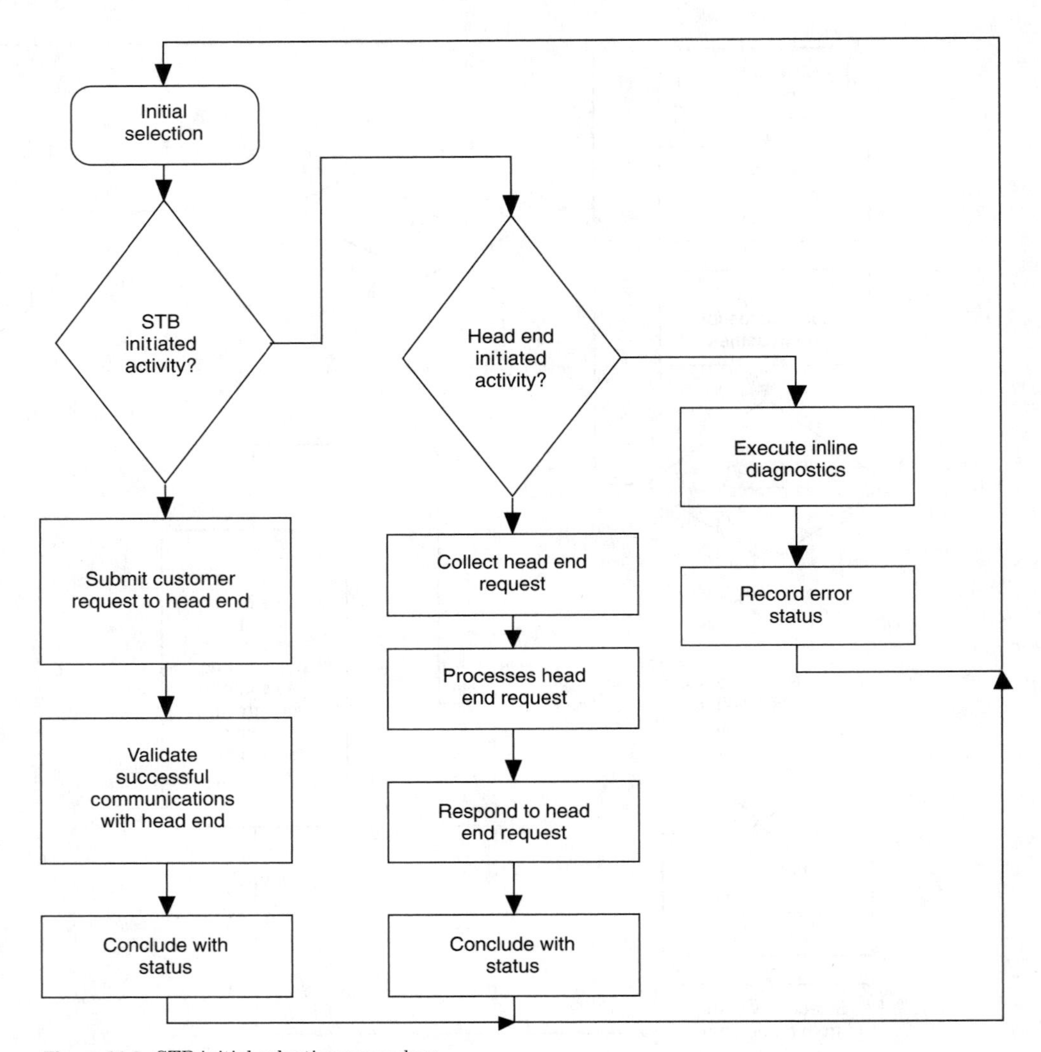

Figure 11.8 STB initial selection procedure.

Regardless of the type of remote being used, the type of menu is the same. The menu selections each have a number and text describing the menu bar selection. A menu item may be selected either by entering the one- or two-digit code identifying it, or by targeting it with the laser dot emitted by the one-button control and clicking on it. In either case, the appropriate software decodes the user action into a request to be sent to the video server. This could be a request for another menu level, a program selection, or an order for a product being displayed for sale.

In summary, this section has described the functionality of a low-cost set top box that can be used to interface either to a CATV coaxial cable, or to conventional twisted-pair telephone lines. An open architecture was proposed that will:

- Permit interfacing to different video servers via the CATV network
- Facilitate different decompression algorithms to be used for each program, or even for each scene
- Enable system and comprehensive network and total system analysis never before possible

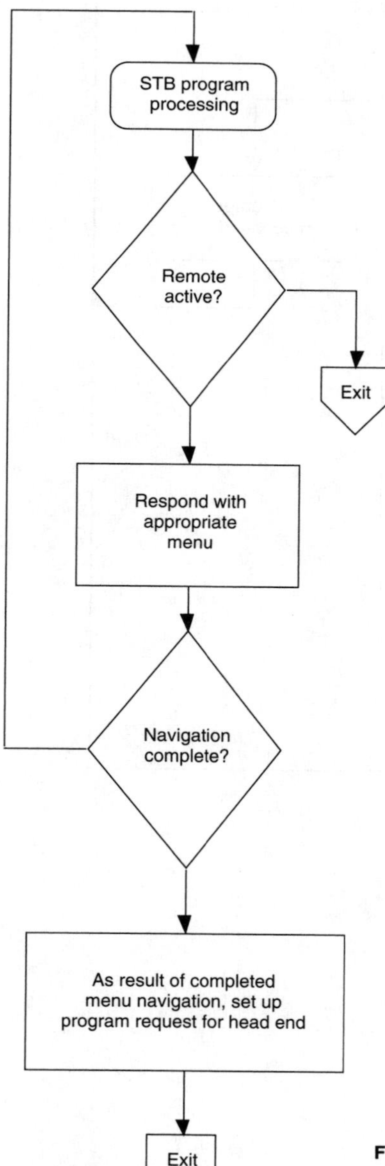

Figure 11.9 STB program processing.

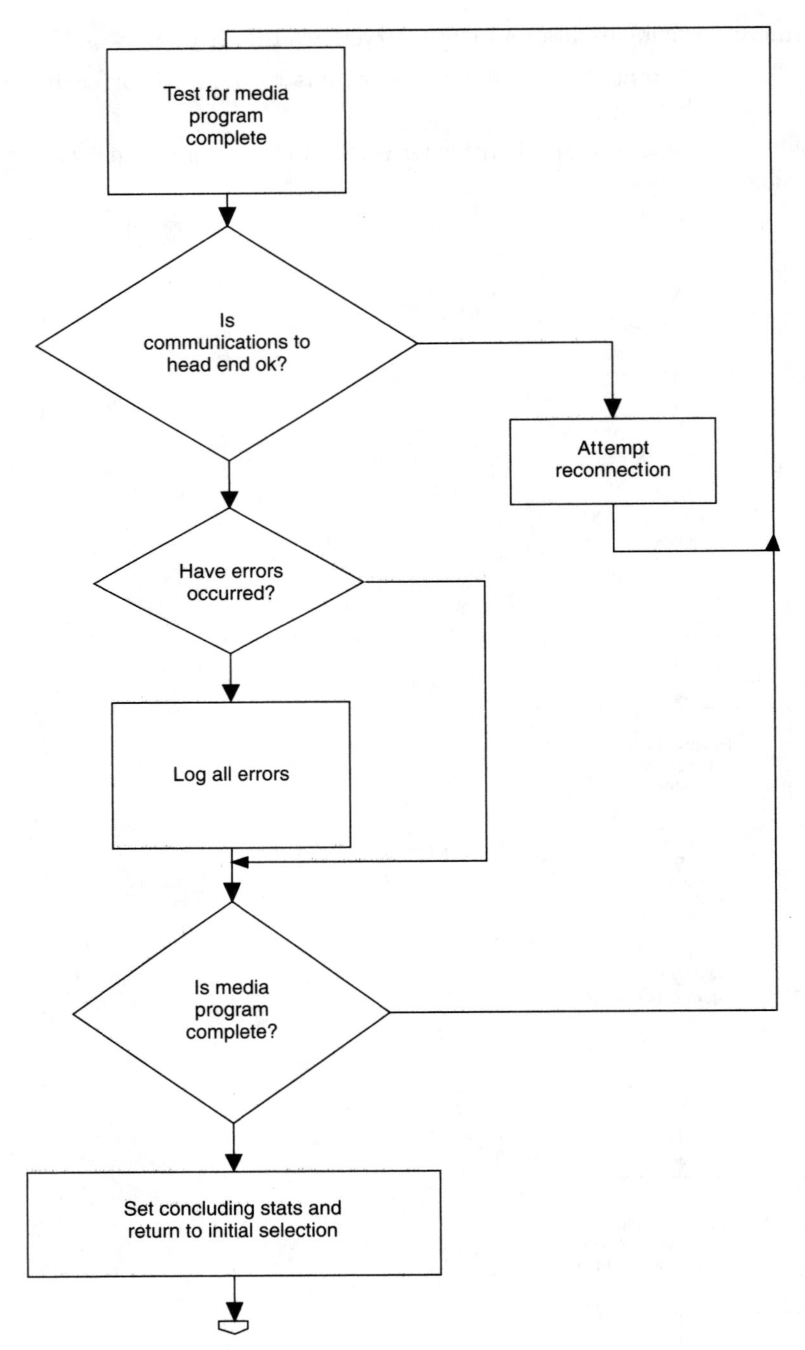

Figure 11.10 STB program completion.

The open architecture also allows for the incorporation of future algorithms as they are developed and improved. An eventual cost target of $150 for the basic STB was proposed. Sample program flows were presented, and the characteristics and control philosophies for two different types of remote controls were discussed.

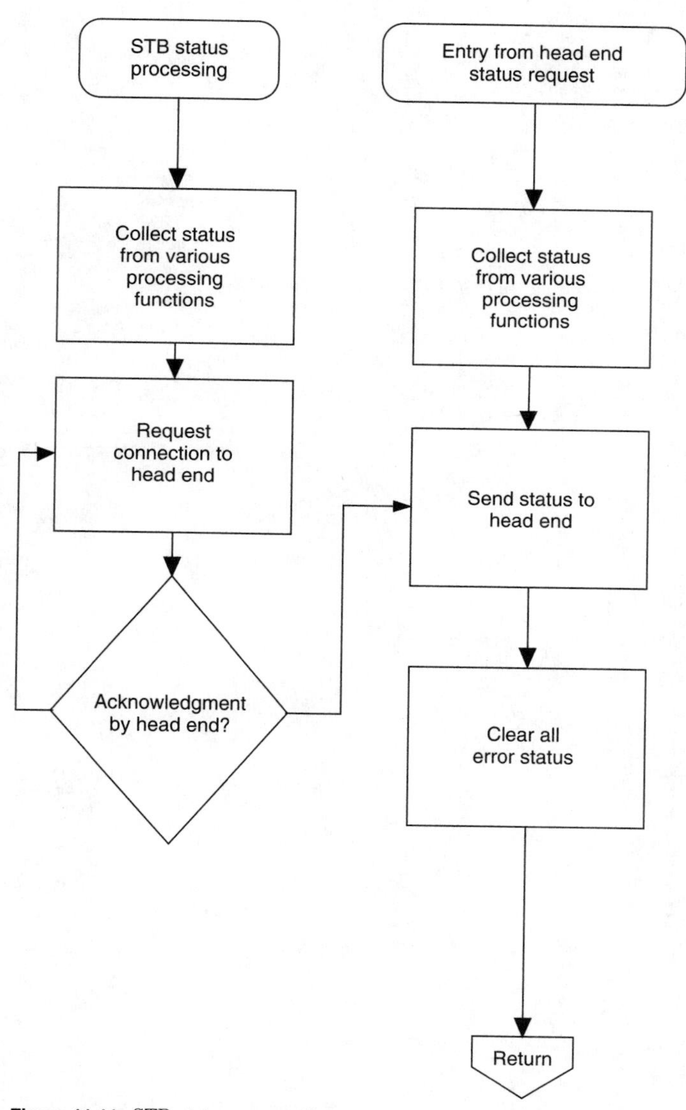

Figure 11.11 STB status processing.

Chapter

12

System Value Engineering Requirements

The converging technologies of ITV and advanced multimedia (AMM) have created a single application that rivals in scope and magnitude all recent engineering events. Millions of computer users and many millions of TV viewers will be affected by this technology convergence revolution. Video servers, networks, switching systems, and user interfaces will be replicated on a scale that, if done correctly, could consolidate and revolutionize the entire world. Common standards, user interfaces, and hardware will permit very significant mass production. This mass production, if managed properly, would lead to economies of scale that would ultimately permit the inclusion of the world's entire population to this event.

The ITV and AMM application would absorb millions of video servers, hundreds of millions of ITV set top boxes, tens of millions of computer interfaces, and humongous amounts of networking facilities. This is truly a multibillion dollar endeavor. It is therefore important that the architecture of such an endeavor not be fixed nor closed, because this technology should be permitted to evolve with open standards to continue to meet the continuing new requirements of its users.

The concept of using set top boxes whose characteristics can be established by ROM or downloaded to RAM is important. The ability to dynamically improve communications, compression, and control will ensure that this ITV/AMM technology will have a long and productive life.

We have seen in the past that programmable, generalized hardware design applied to all functional electronic building blocks does not increase overall system cost. Frequently it reduces the system cost and cost of ownership by permitting the inclusion of inexpensive programmable diagnostic facilities without added hardware.

If the video servers and switches are designed to be CATV network and set top box-friendly, and if the STB is likewise network- and video server-friendly, then the assurance that this technology can be applied with equal ease to CATV, TELCO, hotel PPV, and aircraft entertainment applications can be ensured, and the evolution can continue.

Creative architecture and system design will permit cost-effective implementation, maintenance, and evolution. Technology obsolescence will be postponed. Perhaps with this technology approach, 50+ years of utilization may be extracted from it, as was the case for NTSC TV.

Chapter

13

A Film Quality Digital Archiving and Editing System

Note: This chapter was presented as a paper at the "Advance Television & Electronic Imaging for Video & Film," SMPTE Conference Feb. 5-6, 1993, held in New York City and delivered by Winston Hodge (Hodge Computer Research Corp.), Robert S. Block (International Communication Technologies, Inc.), and Bill Harvey (Next Century Media, Inc.)

Film archiving of commercial theatrical motion picture film has not been as successful as was originally expected. Continuing degradation of film quality over time has represented serious loss to the major motion picture studios. This article is a proposal to rectify these deficiencies using new digital technology with advanced state-of-the-art image processing and database technology. Furthermore, because of the similarity of a digital archiving and very high-resolution editing system, consolidation of those features is discussed.

More than 50% of the Hollywood motion picture films produced before 1950 are no longer in existence. The loss grows to more than 80% of Hollywood movies if one considers films produced before 1930. Within one lifetime these records have literally disintegrated. And that is just the tip of the iceberg. Hollywood films are among the most commercially valuable pictures. Their preservation record shines compared to that of other films. The losses are even greater for many other historically important and commercially valuable films. This devastating look into the past does not change much when we look into the future. There is still no way to preserve film permanently. Eastman Kodak's recent announcement of the molecular sieve technology provides for the prevention of the "vinegar syndrome" in acetate film. However, once the vinegar syndrome begins, the molecular sieve can do nothing to stop it, Kodak admits, and it is estimated that most of the acetate films in existence have already gone into vinegar syndrome. Therefore the molecular sieve can basically only protect the new films made today.

Oscar-winning cinematographer Alan Daviau (*E.T.*, *Bugsy*, etc.) argues that the "YCM" method of preservation used today simply does not allow one to return to a *high-quality* print, and that "This is the emperor's new clothes of the business." Therefore, despite the molecular sieve, no method of film preservation exists that has been demonstrated to solve the physical decay problem and the artistic criteria of the industry.

More than that, much of the valuable footage produced in the past 20 years has been produced on magnetic videotape. Its life span is far less than that of film. There is no record of videotape loss, but the authors believe that more than 50% of the videotape produced as recently as 10 years ago may be lost. It is disturbing to think of the cultural, scientific, educational, historical, sport, and entertainment film that will be lost to future generations.

This article explores a practical, cost-effective method to overcome this problem using advanced, state-of-the-art digital image processing and database technology. Furthermore, once very high-resolution digital conversion is accomplished, there are many other aspects of film making that can benefit from digital image processing. These include editing, special effects, and the distribution of motion pictures in a digital format for improved visual and audio performance.

The Problem

Storing film in a digital format requires a very large storage capacity. To maintain 35mm film quality and resolution, a theatrical-release motion picture film must be sampled at a rate of 3072 × 4096 picture elements (pixels) per 35mm frame. Each pixel requires approximately 24-bit (some people assume 32-bit) color, and that is 37.7 megabits per frame. Table 13.1 below extrapolates the data requirements per 1 hour of film to be 3.261 terabytes, and this does not include sound tracks or other pertinent data.

TABLE 13.1 Image Data Requirements

Frame Rate	
Frames per second	24
Frames per minute	1,440
Frames per hour	86,400
35mm Film Resolution	
Horizontal pixels:	4,096
Vertical pixels:	3,072
Area (pixels per 35mm frame):	12,582,912
Bits required using 24-bit color pixels:	301,989,888
Number of bytes per 35mm frame:	37,748,736
Storage bytes required for a one-hour color film:	3,261,490,790,400
Hours of films to archive:	10,000
Total storage requirements in bytes:	32,614,907,904,000,000

TABLE 13.2 Assumptions Used to Determine Raw Acoustic Requirements

Sample rate (samples/second)	44,000
Amplitude resolution in bits	16
Bits per second per channel	704,000
Number of channels	4
Aggregate bit rate per second	2,816,000
Aggregate bit rate per minute	168,960,000
Aggregate bit rate per hour	10,137,600,000
Aggregate acoustic byte rate per hour	1,267,200,000
Image hour byte rate (uncompressed)	3,261,490,790,400
Percent audio to video (uncompressed)	0.04%

A digitized sound track can be shown only to add about .004% more data to a digitized film, assuming 4 channels of CD-quality sound. This is represented by Table 13.2 see above.

The Solution

The computer industry has understood information preservation technology for years. The problem of applying computer technology to film preservation has been, until recently, a problem of data storage capacity. A data processing facility with a capacity to manage, manipulate, and validate thousands of terabytes of storage is almost unheard of.

1 terabyte = 1000 gigabytes.
1 gigabyte = 1000 megabytes.
1 megabyte = 1000 kilobytes.
1 kilobyte = 1000 bytes

Recently, however, random access and sequential access storage technology has become available with multiple-terabyte capacities. Within 18 months, on-line data storage capacities will reach thousands of terabytes. The time when we have the capacity to digitally store and archive film is arriving.

Figure 13.1 below depicts realizable storage capacities as a function of time. The smooth curve is exponential, approximating storage capacity expectations into the year 2000. From this we can see that data storage capacity will not be the limiting factor to archiving motion picture film.

It is well understood that digital storage and transmission technology can be as fault-tolerant as desired to preserve data integrity by assigning error correction codes to blocks of data. Frequent examination of these data blocks will reveal their level of data integrity. The error correction codes will determine if a bit or block of errors has occurred and, if so, will correct them. These error correction codes require appendages (or syndromes) to be added to each block. This increases the raw data requirement fur-

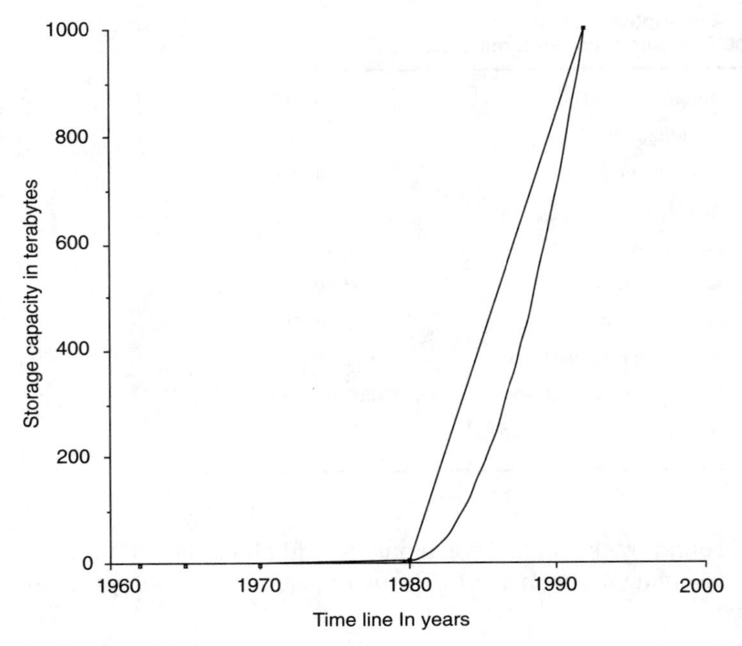

Figure 13.1 Storage capacity vs. time.

ther, but only a minuscule amount. Because magnetic media (as film media) will degrade as a function of time, a digital film archiving system will need to be validated periodically. The technology for validating digital data is well known and has been applied throughout the computer industry for the last 25 years. The secret is to validate and correct long before it would become impossible to do so. This means a digitized film archiving system would require a computing system to access and validate film data.

Required Computer Performance

Looking at Table 13.1, note that each single frame requires *37,748,736* bytes for image representation. For a computer merely to pass this data through its system at a rate of 24 frames per second (real time) would require a clock rate of over 960 MHz. To do any kind of computation, including validation, image processing, etc., would require many times this clock rate, along with extensive data addressing capability. Obviously, this is both a storage-intensive and computer performance-intensive problem. Computer-intensive problems can often be solved by applying multiple parallel processors to the problem until performance expectations are realized, by tailoring and optimizing computing algorithms, and by representing data in more compact form (compression).

System Requirements

Figure 13.2, represents a high-level view of a film archiving system.

A very high-resolution film editing facility has been incorporated in the system design because only a modest increase in hardware cost is involved.

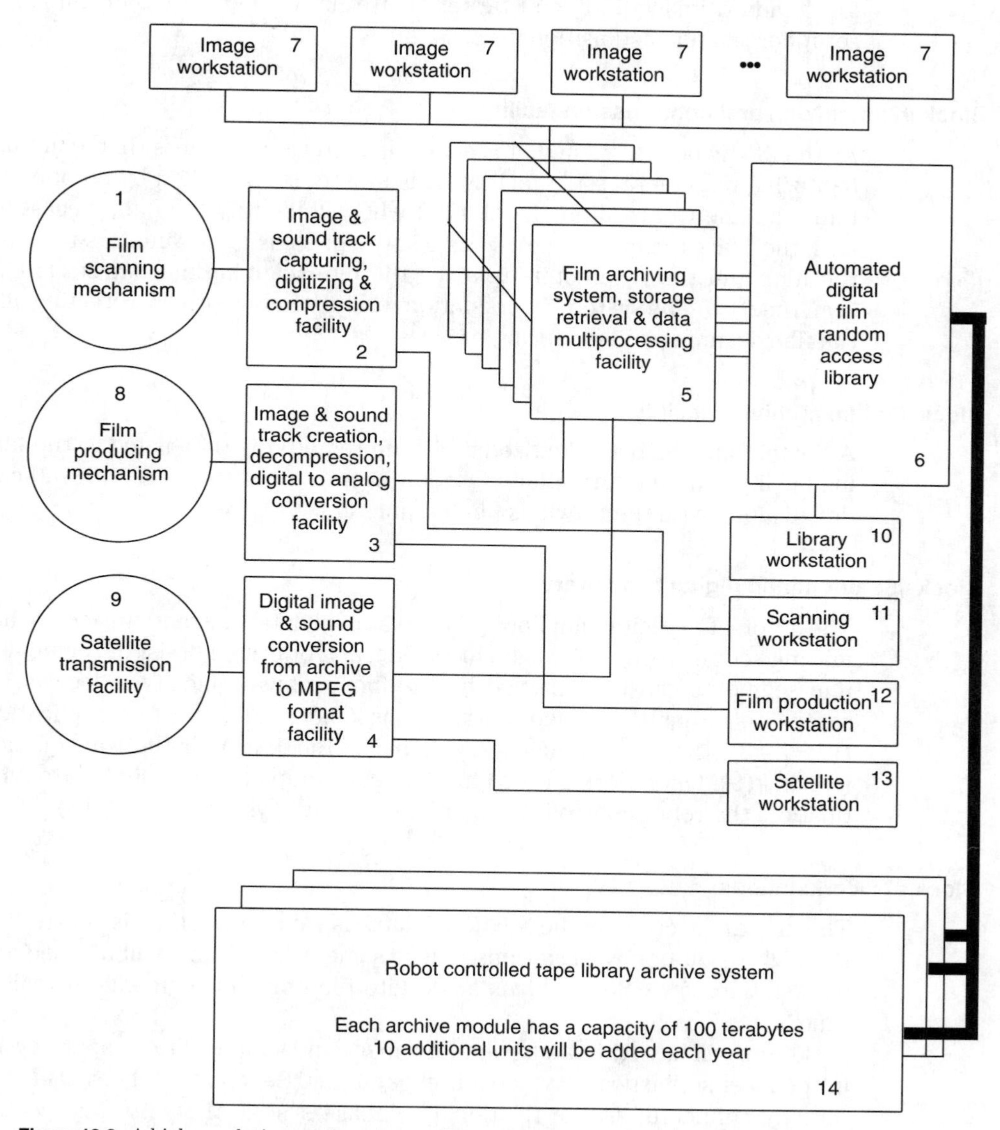

Figure 13.2 A high-resolution motion picture film archiving and editing system.

Block #1: film scanning mechanism

This is where the initial acquisition of film image and sound track starts. The highest-grade studio film negatives or positives are fed to the high-resolution film archiving system. The images are scanned and digitized, and the sound tracks are converted to CD-quality multitrack digital audio. The film scanning device should have automated or simplified mechanisms providing for validation of image colors, focus, variations in transfer function etc., as well as sound track validation. Facilities for standard image and acoustic test patterns will permit easier initialization and setup for the film acquisition process.

Block #2: digitizing and compression facility

As the film is being scanned, it is also being fed (at a rate less than real time) to the digitizing and compression facility. This system is controlled by an operator at the film scanning workstation (Block #11) who will identify the film, its cast of characters, the film's technical characteristics, the processing parameters used to process the film, and the parameters used by the film scanning mechanism (Block #1) to scan the data. The system's workstation will be window-driven, directing all facets of the film scanning and capturing process.

Block #5: film archiving facility

After the film has been digitized and compressed, the data is fed to the film archiving facility, where information regarding scenes, producer, director, and cast are indexed along with the previous film parameters.

Block #6: automated digital film library

The automated digital film library is a relational database random access facility permitting access to any film segment (scene, act, or other) or list of segments. A list of film segments can be retrieved by the film archiving facility (Block #5) and presented to either the image workstations (Block #7), the film replication facility (Block #3), the film producing mechanism (Block #8), or the transmission facility (Block #(9). Once work on the film has been completed, it is sent on to its final destination, the robot-controlled tape library archive system (Block #14).

Block #7: workstations

The process of creating the relational database at archival time is controlled through the high-resolution workstations. These same workstations will be used as film edit workstations for color or black-and-white film, or film segments in real or dilated time.

Editors will have the same complete image and sound editing capability as is found in the most sophisticated video editing systems. Beyond that, because the system is directly connected to the relational database system the editor can watch a completed film on his or her image workstation, search for specific scenes from a complete database of films, combine pieces from various archived image sources, and alter sequences, colors, rates, contents, fades, dissolves, rotations, stabilization, etc.

The image workstations can combine digital special effects and graphics with film segments (titling, graphic overlays, etc.). Foregrounds, backgrounds, portions of scenes, or whole scenes from one film can be easily and seamlessly inserted into another film. Editors will find their creative latitude greatly expanded with this system.

Often films are edited differently for different audiences. Thus, there may be more than one version of the film. To accommodate this requirement, multiple edits, based on different film segment lists (linking lists), can be stored and selected in the automated digital film library (Block #6) and the remote controlled library archive system (Block #14). There will likely be multiple linking lists for each film. The film production workstation (Block #12) or the workstation (Block #13) merely requests the film, the desired linking list, time dilation for each element of the linking list, commercial allocation time if any, etc. These workstations can also schedule work in advance for several months to assure high productivity.

Block #3: film replication facility

This is where the digitized film is converted to a format usable by the film producing mechanism (Block #8).

Block #8: film producing mechanism

This block includes decompression, digital-to-analog conversion, and presentation to the film duplication equipment. Provisions will be made to transmit retrieved digitized films for presentation to network and cable TV head ends in multiple formats, including MPEG, NTSC, SECAM, PAL, and HDTV.

This function will be performed by the digital and sound format conversion facility (Block #4). The processed data is then fed to the transmission facility (Block #9) for transmission, or to the film producing mechanism (Block #8) to be recorded on videotape.

System Operation: Data Flow

Color positive or negative film will be scanned by the film scanning mechanism (Block #1). Because color negative film has much higher resolution than positive film, and positive film has much higher contrast, facilities will be provided to preserve the maximum film information content. Therefore, image capturing for negative and positive film should require somewhat different processing to accommodate variations in transfer function (dynamic range, contrast, gamma factor, grain, etc.). Image compression algorithms should be applied at this time to minimize data transfer time and storage (Block #2).

As the film images are being digitized, the sound track is similarly being processed. Up to 7 (usually 4) channels of acoustic information will be digitized and combined on the same media as the film descriptors and images (Block #2). Because it will be necessary to slow the film down (*time dilation*) on a viewing workstation without affecting the tones of the actors' voices, music, etc., appropriate sampling, digitizing, and formatting must be done at this phase. Acoustic data compression can also be applied at this phase.

The digitized film media (processed by Block #2) will be formatted, identifying the film, its date of copyright, the list of actors and producers, a brief summary, technical characteristics of film, a list of scenes, and so forth. This information can be entered from the attached scanning workstation (Block #11).

Time dilation on the editing workstations (Block #7) should similarly not degrade the displayed image sequences. Time dilation applied to all or part of a film shall not adversely affect the image quality of the film displayed by the film replication facility, nor shall the transmission or recording on videotape degrade the film below the standards for each format used.

A fully automated storage hierarchy is created between Block #6, the automated random access library, and Block #14, the robot-controlled tape library archive system. Processors embedded within these two units communicate and cause any multiples of terabytes to be downloaded from Block #14 to Block #6 and vice-versa. Block #4 could be a silo manufactured by Storage Technology Corporation, each silo providing on-line storage to 100 terabytes of data. Multiple silos could be attached. The contents of the data on silos is validated and revalidated frequently by internal subsystem processors. Errors will be logged and corrected.

The film capturing and compression facility (Block #2) will be electrically connected to the mainframe responsible for film archiving and retrieval. Electrically and logically, the film compression facility shall have a standard appearance to the mainframe, such as that of a serial communications device, tape subsystem, SCSI, or other interface. The concept of using standard interfaces between major subsystem blocks makes for easier maintenance, replacement of failing major blocks, debugging, upgrading, cabling, etc.

The film archiving facility (Block #5) shall be responsible for reformatting general text data from the Film Capturing Facility and reformatting and creating a relational database of an IBM DB2, Oracle or Ingress vintage. The Film Archiving Facility shall be responsible for controlling the Automated Digital Film Library, which is its peripheral.

The automated digital film library (Block #6) may be an optical or magnetic storage device (probably a large group of them) with capacity to store thousands of terabytes of digitized film data. This system could almost be an existing fully functioning product. It is expected that the automated library system will occupy a very large physical space.

The culmination process will be the conversion of digital film data back to film (Block #8). Image and acoustic time dilation shall be facilitated both to the editing workstations and the film replication facility. The system must be a digital imaging system permitting 35mm theatrical release motion picture film to be scanned, digitized, compressed, edited, archived, and reprinted on motion picture film without *perceivable degradation* as compared with the existing procedures using film negatives and positives. Without perceivable degradation means that no change of quality or character is discernible to the naked eye of a trained and skilled observer when viewed on a large theater screen. It does not mean that all pixel values of the original film will have the exact numerical value as the replicated film. Of course, even film-to-film copies are not exactly the same at the pixel level as the original.

Electronic distribution in a variety of formats is required. This is facilitated by the transmission facility (Block #9) and the film replication facility (Block #8).

Image Data Compression and Formatting

After being scanned by the scanner, digitized film data will be compressed, formatted, and archived into a digital film library. Digital film data will be created and formatted to provide easy database access for digital film editing, manipulation, extraction, and general or selective viewing. Direct access to the data will be achieved through high-performance graphics workstations. Lossless or virtually lossless compression that does not create human-observable artifacts can permit significant reduction in archiving and processing costs.

While the investment in hardware and software is examined, it becomes obvious that other relevant and salient features can be added with a very small cost increment. These features include electronic distribution and very high-resolution film editing.

Compression of the terabytes of data per film is necessary for numerous reasons, including storage, transmission, database accessing, etc. Compression has both performance and cost consequence. Three types of compression must be considered. They are, according to importance, image compression for compressing the immense amounts of image data, acoustic data compression of the one to seven channels of CD-quality acoustic data, and conventional data compression of frame, scene, and database descriptors.

Compression/decompression algorithm development is independent of the image capture or storage archiving function, but is used by both. We plan to and are currently evaluating image compression algorithms, and we are comparing them to quantify the level of perceptible artifacts they produce. Perceptible artifacts are not acceptable. Artifacts introduced with JPEG and MPEG standards will not be acceptable for this very high-resolution film archiving and editing system.

Lossless or virtually lossless compression will be necessary so there will be no image degradation noticeable to a skilled or trained observer when projected on a large theater screen. The compression shall not exhibit any perceptible degradation of images or acoustic data nor shall it include any discernible artifacts in either the color, spatial, or time domains. Compression factors in the range of 5 times to 50 times are not only desirable, but are believed to be realizable. Algorithms that are reversible (from a critical human perceiver point of view) and simple from a computational point of view are advantageous.

After being scanned by the scanner, digitized film data will be compressed, formatted, and archived into a digital film library. Amortizing system hardware, storage facilities, and media, one can compute the cost of storage for each 7.5-terabyte film. If 7.5 to one compression is assumed, the subsequent numbers are probably realizable. If no compression is provided, then the subsequent costs may be multiplied by 7.5. Thus the economic requirement for compression is portrayed, especially when the compression technology creates *no* film degradation perceptible by a trained observer.

Some hypothetical numbers have been put together to show that for the archiving of 10,000 films, the pro-rata contribution of hardware and storage facilities and the amortization of nonrecurring costs is very small, and the main cost is the media in the robot-controlled tape library archive system. If we assume, for example, that nonmedia costs (including facilities and salaries) cost $50,000,000, then the prorata contribution less media is about $5000 per film, plus profits! The media cost for the robot-controlled tape library archive system may cost under $1500 per film, providing a total archiving cost of about $6500 plus some very small annual archiving and validating fee of perhaps 15% per year. Of course these numbers do not represent cost to the film creators, because they do not include risk funds and profits for the archive service providers.

90-minute film compression summary:

No-compression digital archive costs $19,500
7.5 × compression archive costs $6500
The conclusion: **compression is required.**

After compression, digital film data will be formatted to provide easy database access for digital film editing, manipulation, extraction, and general or selective viewing. Direct access to the data will be achieved through high-performance graphics workstations.

Data validation and automated correction

This digital technology must be self-correcting and error-tolerant, and requires vast amounts of digital storage. Tables 13.1 and 13.2 are shown only to illustrate the general data relationships between image data requirements and acoustic data requirements. Table 13.1 depicts the image storage requirements for one hour of film, not including the acoustic and frame data—one to seven CD-quality voice tracks, frame indexing, scene id, sequential frame number, act id, actor id, producer code, color temperature coding, film data, negative or positive designation, negative or positive processing parameters, error correction codes, etc.

Image and sound capturing will be achieved by a high-performance graphics workstation. The capturing process will include facilities for validating digitizing parameters and to set up characterizing parameters for each scene. Those parameters will be appended as digital descriptors on the digitized media.

Image, acoustic, and data compression are most logically applied at the image and sound capturing facility, and even though it will not be necessary to represent the exact archived digital film format, it shall bear similarities. It is important to compress images and acoustic information as soon as possible because pushing terabytes of data through even a very high-performance workstation requires a large number of computer clock cycles and a corresponding amount of time.

Transmission facility

Electronic distribution to theaters is one of the promises of a centralized high-resolution film archive system. This would permit the highest quality video to be delivered to theater-based electronic image projection systems. Companies such as Hughes are currently working on laser theater-quality image projection systems that are rumored to produce image quality soon to be competitive with film projection systems.

Transmission to motion picture theaters, broadcast networks, cable TV, MMDS, DBS, and other distributors of film information such as universities, libraries, etc., can be accomplished through wideband fiber-optic systems and satellite transmission. The transmission facility should be interconnected with these distribution systems for maximum efficiency.

With a computerized system, each frame of a film can be fully identified so that any film frame image or set of images can be accessed as a relational database or by sequential frame number. This would permit the extraction of film sequences for theater or TV review, or presentation posters.

Library characteristics

Each year, perhaps one thousand films will be digitized, database indexed, and archived (2.74 films per day, working 7 days per week). All films will become part of the database, with the capability of accessing any individual frame from a common image database that could grow at an approximate rate of several kilo-terabytes (KTB) per year, depending on the compression technology utilized.

Storage, archiving, and retrieval of hundreds of KTBs of data is an enormous automated library function. When combined with automation facilities to validate the integrity of the archived data on a regular basis, in order to guarantee film integrity, a very large and high-performance data processing facility will be required. The level of automation must include facilities that automate the storage, retrieval, and handling of the storage media, and of course, all films will be revalidated periodically and all errors will be detected and corrected.

Database functions

In order to archive the data and preserve it immaculately for all time, vast storage and very significant computer power are not all that is required. At this point, we have capabilities for a comprehensive relational film database with little additional cost. Without additional hardware or significant additional technology investment, the elements of a film can become part of the database as well as the various films themselves. It would be possible to append to each act and scene relevant information as to who was involved in those elements, what type of film was used, what the transfer functions are, how the film was processed, lapse time of the film element, what type of lighting was employed and what were its characteristics, etc.

It will be the objective of the integrated facility to capture and archive at least three 90- to 100-minute films per day. The films must be retrievable via a standard SQL type database (DB2, Ingress, Oracle, etc.). Digitized films must be able to be edited from an editing workstation using both database commands and editing commands.

Numerous books have been written on video editing, and it is an extensive subject. Video editing technology is well-developed, and the features of video editing must be available to the very high-resolution film editing workstations (7). A list of these features can be found in the book *Video Editing and Post Production: A Professional Guide* by Gary H. Anderson, published by Knowledge Industry Publications. Professional video editing includes sophisticated functions permitting image sequence slowdown or speedup without affecting the image quality or the tonal quality of speech, music, etc.

Very high-resolution editing workstation features should include:

- Formatting
- Digital film editing (very high-resolution)
- Image and sound editing functions
- Password protection to prevent unauthorized edits

- Cut and paste (image and sound)
- Zoom, pan, tilt and rotate, fades, dissolves, wipes
- Colorize and track colorized elements
- Background/foreground differentiation
- Image enhancement/blur control
- Combine elements
- Add elements (titles, images, etc.)
- Remove elements
- Image stabilization, vibration removal tools
- Special effects, paint, saturation control, etc.
- Assemble and insert edits, image sequence selection, start codes, stop codes
- Edit lists, list creation, review, modify and execute
- Transfer function modification (all or part of image)
- SMPTE test code validation/normalization
- Synchronization
- Time code and frame code editing
- Time dilation
- Multitrack audio source mixing, audio edit list, A/V list correlation, and stop/start lists
- Audiovisual time displacement control
- Copy final edit list to new film database
- Morphing

It will be possible to create film hybrids by selecting frames from one part of the composite database and combining them with frames from other parts. Comprehensive cut-and-paste functionality should be provided via attached image workstations.

Conclusions

We have demonstrated that, using existing storage and computational technologies, and even with the worst-case assumption of modest 7× compression, 35mm digital preservation is today realizable at $50,000 per motion picture. This compares with a $35,000 recurring cost for the current method, the "YCM" triple-separation negative process, which must be repeated at 10- to 20-year intervals in most cases (losing a generation of clarity each time), and which has been found to be qualitatively unacceptable by many leading creative people in the motion picture industry. We believe we will demonstrate that a 7.5 to 1 compression will produce no perceivable artifacts. Greater compression factors will also be available to provide a level of archiving that is even more affordable.

Glossary

The glossary definitions below are meant to explain terms used in this text and industry. There exist more formal definitions for many of these terms in standards documents and other, more complete glossaries. The purpose of this glossary is to promote rapid understanding and explanation. For formal and exact definitions, the reader should refer to formal standards and glossaries such as those published by the NCTA, EIA, CCITT, and other standards bodies.

AAR Automatic Alternate Routing (in a telephone switching system).

access unit [system] In the case of compressed audio an access unit is an audio access unit. In the case of compressed video, an access unit is the coded representation of a picture.

ac coefficient Any DCT coefficient for which the frequency in one or both dimensions is nonzero.

adaptive bit allocation [audio] The assignment of bits to subbands in a time and frequency varying fashion according to a psychoacoustic model.

adaptive multichannel prediction [audio] A method of multichannel data reduction exploiting statistical interchannel dependencies.

adaptive noise allocation [audio] The assignment of coding noise to frequency bands in a time and frequency varying fashion according to a psychoacoustic model.

adaptive segmentation [audio] A subdivision of the digital representation of an audio signal in variable segments of time.

address An indication of the location of the intended recipient or storage location for a message.

addressing Transmission in which the location or identity of intended recipient(s) is indicated.

ADSL (Asynchronous digital subscriber loop) Developed at Bellcore, this compression technology permits the delivery of video through copper wire. Pacific Bell views ADSL as a transitory technology for niche applications with only a few years of economic life.

advanced multimedia a smoother, higher-resolution, more provocative, higher-fidelity form of multimedia. See multimedia.

aggregate A single bit stream combining many bit streams.

alias [audio] Mirrored signal component resulting from sub-Nyquist sampling.

alphabet The collection of symbols in a code.

alternating current Periodically reversing electric current.

analog Characterized by a continuous (vs. discrete) range of values.

analog transmission The traditional telephone technology in which sound waves or other information are converted into electrical impulses of varying strengths. Most cable programming available today is sent to customer homes in analog format.

analysis filter bank [audio] Filter bank in the encoder that transforms a broadband PCM audio signal into a set of subsampled subband samples.

ANSI American National Standards Institute.

ASCII American Standard Code for Information Interchange.

ASDL Asynchronous Subscriber Data Link, a technology developed expressly for TELCOs that permits POTS (Plain Old Telephone Service), ISDN, and 1.5- to 6-Mbps video data over several miles of conventional TELCO twisted pair, to facilitate ITV.

assigned frequency The frequency coinciding with the center of the radio frequency channel in which the station is authorized to work.

Asynchronous Transfer Mode See ATM.

asynchronous Without regular intersymbol timing.

ATM (Asynchronous Transfer Mode) An ultra-high-speed switching technology that can simultaneously route voice, data, and video communications over fiber-optic lines at speeds eventually reaching billions of bits per second. Pacific Bell plans to install ATM switching machines in each of its serving areas. This architecture directly supports the TELCO Video Dial Tone concept and is synergistic with CATV ATM deployment strategy.

attenuation Loss or diminution, usually of transmitted signal.

audio access unit [audio] For layers 1 and 2, an audio access unit is defined as the smallest part of the encoded bit stream that can be decoded by itself, where decoded means "fully reconstructed sound." For layer 3, an audio access unit is part of the bit stream that is decodable with the use of previously acquired main information.

audio buffer [audio] A buffer in the system target decoder for storage of compressed audio data.

audio sequence [audio] A noninterrupted series of audio frames in which the ID, layer, sampling frequency, and layer 1 and 2 bit-rate index parameters are not changed.

AWG American Wire Gauge (size of electrical wires).

B-field picture A field structure B-picture.

B-frame picture A frame structure B-picture.

B-picture (bidirectionally predictive-coded picture) A picture that is coded using motion compensated prediction from past and/or future reference fields or frames.

backward compatibility A newer coding standard is backward-compatible with an older coding standard if decoders designed to operate with the older coding standards are able to continue to operate by decoding all or part of a bit stream produced according to the newer coding standard.

backward motion vector [video] A motion vector that is used for motion compensation from a reference picture at a later time in display order.

bandwidth A measurement of transmission capacity; the greater the bandwidth, the greater the information-carrying capability of the transmission medium. Analog transmission is measured in cycles per second. Digital transmission is measured in bits of information per second.

bark [audio] Unit of critical band rate. The Bark scale is a nonlinear mapping of the frequency scale over the audio range, closely corresponding with the frequency selectivity of the human ear across the band.

baseband Original and unmodulated information frequency band.

baud Unit of signal rate.

bidirectionality predictive-coded picture; B-picture [video] A picture that is coded using motion-compensated prediction from a past and/or future reference picture.

BISDN Broadband Integrated Services Digital Network.

bit rate The rate at which the coded bit stream is delivered from the storage medium to the input of a decoder.

block An 8-row by 8-column matrix of samples, or 64 DCT coefficients (source, quantized, or dequantized).

block companding [audio] Normalizing of the digital representation of an audio signal within a certain time period.

block [video] An 8-row-by-8-column orthogonal block of pixels.

bottom field One of two fields that comprise a frame. Each line of a bottom field is spatially located immediately below the corresponding line of the top field.

bound [audio] The lowest subband in which intensity stereo coding is used.

broadband communications Communications that require extremely high levels of transmission capacity (or bandwidth). For example, video transmission and interactive services require the ability to send very large amounts of information down the pipeline, while voice communications require transmission of relatively small amounts of information. Fiber-optic and coaxial cable can carry broadband communications; the copper wires traditionally used by telephone companies can't, unless advanced compression techniques are used.

broadband network A network capable of transporting voice, interactive full-motion video, and data services. Technically, a network capable of transmitting 1.5 million or more pieces of information per second. A narrowband network carries significantly less information than a broadband network. Narrowband applications include traditional telephone service, electronic mail, paging services, and faxes. Technically, a narrowband network carries up to 64,000 pieces of information per second. A wideband network is capable of carrying less information than a broadband network, but more than a narrowband network. Services possible over a wideband network include video teleconferencing, file transfer, and video telephony. Technically, wideband services are those that require between 64,000 and 1.5 million bits of information per second.

broadcast video Video whose quality is about equal to a transmission from a television station to a receiver with a good receiving antenna at a distance of about 2 miles. Good quality television. The digital MPEG 2 equivalent is considered about 5 Mbps, compared to VHS quality, which is considered to be about 1.5 Mbps.

brouter Combination bridge and routing devices to interconnect LANs and or WANs.

byte Sequence of 8 bits.

byte aligned A bit in a coded bit stream is byte-aligned if its position is a multiple of 8 bits from the first bit in the stream.

carrier (signal or wave) In a frequency stabilized system, the sinusoidal component of a modulated wave; the output of a transmitter when the modulating wave is made zero; a wave generated at a point in the transmitting system and subsequently modulated by the signal; a wave generated locally at the receiving terminal that, when combined with the sidebands in a suitable detector, produces the modulating wave.

carrier frequency The frequency of a carrier (signal or wave).

carrier system A method of carrying several information channels.

CATV Abbreviation for cable television.

center channel [audio] An audio presentation channel used to stabilize the central component of the frontal stereo image.

central office A land-line termination center used for switching and interconnection of public message communications circuits.

channel A digital medium that stores or transports a bit stream constructed according to a specification.

channel [audio] A sequence of data representing an audio signal being transported.

channel bank A device to place multiple channels on a digital or analog carrier.

channelizaton Organizing into channels.

channel slot A position of a channel on a carrier.

chroma simulcast A type of scalability (which is a subset of SNR scalability) where the enhancement layers contain only coded refinement data for the dc coefficients, and all the data for the ac coefficients, of the chrominance components.

chrominance (component) A matrix, block, or single sample representing one of the two color difference signals related to the primary colors in the manner defined in the bit stream. The symbols used for the chrominance signals are Cr and Cb.

chrominance format Defines the number of chrominance blocks in a macroblock.

clear channel A channel without any format restrictions.

coaxial cable A transmission line with a central core that conducts electricity (a conductor). This core is surrounded by insulation, followed by another layer of conducting material. Coaxial cable, which is used in the cable industry, can transmit more information than a regular pair of twisted copper wires.

CODEC COder/DECoder, a device to encode and decode signals.

coded audio bit stream [audio] A coded representation of an audio signal as specified in this part of the CD.

coded B-frame A B-frame picture or a pair of B-field pictures.

coded frame A coded frame is a coded I-frame, a coded P-frame, or a coded B-frame.

coded I-frame An I-frame picture or a pair of field pictures, where the first one is an I-picture and the second one is an I-picture or a P-picture.

coded order [video] The order in which the pictures are stored and decoded. This order is not necessarily the same as the display order.

coded P-frame A P-frame picture or a pair of P-field pictures.

coded picture A coded picture is made of a picture header, the optional extensions immediately following it, and the following picture data. A coded picture may be a frame picture or a field picture.

coded representation A data element as represented in its encoded form.

coded video bit stream A coded representation of a series of one or more pictures.

coding parameters The set of user-definable parameters that characterize a coded video bit stream. Bit streams are characterized by coding parameters. Decoders are characterized by the bit streams that they are capable of decoding.

common control An arrangement by which a single control mechanism, such as a stored-program computer, is shared over time by multiple connections.

common equipment Equipment that is shared and not duplicated on a channel-by-channel basis.

communications common carrier Any person engaged in rendering communications service for hire to the public.

companding Nonlinear reduction of a signal, usually compressing the larger signal values.

component A matrix, block, or single sample from one of the three matrices (luminance and two chrominance) that make up a picture.

compression Reduction in the number of bits used to represent a quantity of data.

constant bit rate Operation where the bit rate is constant from start to finish of the coded bit stream.

constant bit rate coded video A compressed video bit stream with a constant average bit rate.

convergence A term popularly used to describe the merging technologies of television, computers, and multimedia.

CRC Cyclic redundancy check.

critical band [audio] Psychoacoustic measure in the spectral domain that corresponds to the frequency selectivity of the human ear. This selectivity is expressed in Barks.

critical band rate [audio] Psychoacoustic function of frequency. At a given audible frequency, it is proportional to the number of critical bands below that frequency. The units of the critical band rate scale are Barks.

cross-connect A place or time where channels (often DS-0 channels) are interconnected.

crosstalk Interference between channels, often on an analog carrier system.

CSDC Circuit-Switched Digital Capability, a service that provides switched digital connections between customers, typically at 56 kbits/s).

CSMA Carrier Sense/Multiple Access, a method of placing information on a common medium.

D-picture A type of picture that shall not be used except in ISO/IEC 11172-2.

data element An item of data as represented before encoding and after decoding.

data partitioning A method for dividing a bit stream into two separate bit streams for error resilience purposes. The two bit streams have to be recombined before decoding.

DB2 A database query language created by IBM. DB2 was the forerunner of the SQL database language.

dc coefficient The DCT coefficient for which the frequency is zero in both dimensions.

DCT Discrete Cosine Transform.

DCT coefficient The amplitude of a specific cosine basis function.

DEC Digital Equipment Corporation.

decibel A logarithmic unit of magnitude comparison named after Alexander Graham Bell, and abbreviated dB.

decode To reverse the effect of coding and return data to an uncoded state.

decoded stream The decoded reconstruction of a compressed bit stream.

decoder Hardware or software that performs the decoding process.

decoder input buffer The first-in first-out (FIFO) buffer specified in the video buffering verifier.

decoder input rate [video] The data rate specified in the video buffering verifier and encoded in the coded video bit stream.

decoding (process) The process defined in this book that reads an input coded bit stream and produces decoded pictures or audio samples.

decoding time-stamp; DTS [system] A field that may be present in a packet header that indicates the time that an access unit is decoded in the system target decoder.

de-emphasis [audio] Filtering applied to an audio signal after storage or transmission to undo a linear distortion due to emphasis.

demarcation A connecting point separating responsibilities.

demodulation The process of extracting intelligence from a modulated signal.

demultiplex To reverse the aggregation effect of multiplexing and return channels to individual states.

dequantization The process of rescaling the quantized DCT coefficients after their representation in the bit stream has been decoded and before they are presented to the inverse DCT.

digital storage media (DSM) A digital storage or transmission device or system.

digital transmission In this form of transmission, sound waves and other information are converted in binary computer code (a series of 0s and 1s). The information is sent in this format, then converted into its original format when it reaches its destination. Digital transmission provides sharper, clearer, and faster transmission than analog transmission, because it can be repeated without introducing noise.

direct broadcast satellite service A radio communication service in which signals transmitted or retransmitted by space stations are intended for direct reception by the general public (includes both individual reception and community reception).

Discrete Cosine Transform (DCT) Either the forward discrete cosine transform or the inverse discrete cosine transform. The DCT is a reversible discrete orthogonal transformation.

display order [video] The order in which the decoded pictures should be displayed. Normally this is the same order in which they were presented at the input of the encoder.

distortion Alteration of a signal in an undesirable way.

down mix [audio] A matrixing of n channels to obtain less than n channels.

downstream The signal path toward the user.

drop cable Coaxial cable that connects to a residence or other service location from a tap on the nearest feeder cable.

dual-channel mode [audio] A mode in which two audio channels with independent program contents (e.g., bilingual) are encoded within one bit stream. The coding process is the same as for the stereo mode.

dynamic crosstalk [audio] A method of multichannel data reduction in which stereo-irrelevant signal components are copied to another channel.

dynamic transmission channel switching [audio] A method of multichannel data reduction by allocating the most orthogonal signal components to the transmission channels.

editing The process by which one or more coded bit streams are manipulated to produce a new coded bit stream.

effective radiated power The product of antenna power input and the antenna power gain (usually expressed in watts or dBW).

emphasis [audio] Filtering applied to an audio signal before storage or transmission to improve the signal-to-noise ratio at high frequencies.

encoder A device or program that performs an encoding process.

encoding (process) A process that reads a stream of input pictures or audio samples and produces a valid coded bit stream.

entropy coding Variable-length lossless coding of the digital representation of a signal to reduce redundancy.

fast-forward playback The process of displaying a sequence (or parts of a sequence) of pictures in display order faster than real-time.

fast-reverse playback The process of displaying the picture sequence in the reverse of display order, faster than real-time.

FAX Abbreviation for facsimile.

feeder cable Coaxial cables that run down the streets or rear easements in the served area and which carry the RF signal from a bridge amplifier on the trunk cable. Bridges might be stand-alone units or be combined in a bridge amplifier. Taps, which are directional couplers, are placed periodically along and in series with the feeder cable, in order to provide service connection points for drops. Losses in the feeder system are made up by line extenders, which are designed for gain rather than for low noise ("high level"). Typically there are at most 2 line extenders in any feeder.

FFT Fast Fourier Transformation. A fast algorithm for performing a discrete Fourier transform (an orthogonal transform).

fiber node size Fiber-optic cable feeders terminate at a node in each distribution area, which serves less than 500 homes. A coaxial distribution network using limited active electronics provides connectivity the rest of the way from the fiber node to the home.

fiber optics A technology that uses light to transport information from one point to another. Fiber-optic cable consists of thin filaments of glass through which light can travel for relatively long distances without the need for amplification. Fiber optics can transmit many more messages simultaneously at higher speeds, with less interference and at lower cost, than other transmission media such as coaxial cable or twisted pairs of copper wire. Some versions of fiber optics now carry 30,000 times as much information as can be carried over copper wire.

field For an interlaced video signal, a "field" is the assembly of alternate lines of a frame. Therefore, an interlaced frame is composed of two fields, a top field and a bottom field.

field period The reciprocal of twice the frame rate.

field picture; field structure picture A field structure picture is a coded picture with a picture structure equal to "top field" or "bottom field."

filter bank [audio] A set of band-pass filters covering the entire audio frequency range.

fixed segmentation [audio] A subdivision of the digital representation of an audio signal into fixed segments of time.

flag A variable that can take one of only the two values defined in this specification.

forbidden The term "forbidden," when used in the clauses defining the coded bit stream, indicates that the value will never be used.

forced updating The process by which macroblocks are intra-coded from time-to-time to ensure that mismatch errors between the inverse DCT processes in encoders and decoders cannot build up excessively.

forward compatibility An older coding standard is forward-compatible with a newer coding standard if decoders designed to operate with the newer coding standard are able to decode bit streams of the older coding standard.

forward motion vector A motion vector that is used for motion compensation from a reference frame or reference field at an earlier time in display order.

frame [audio] A part of the audio signal that corresponds to audio PCM samples from an audio access unit.

frame [video] A frame contains lines of spatial information of a video signal. For progressive video, these lines contain samples starting from one time instant and continuing through successive lines to the bottom of the frame. For interlaced video a frame consists of two fields, a top field and a bottom field. One of these fields will commence one field period later than the other.

frame period The reciprocal of the frame rate.

frame picture; frame structure picture A frame structure picture is a coded picture with picture structure equal to "frame."

frame rate The rate at which frames are output from the decoding process.

free format [audio] Any bit rate other than the defined bit rates that is less than the maximum valid bit rate for each layer.

FSN Full Service Network.

future reference frame (field) A future reference frame (field) is a reference frame (field) that occurs at a later time than the current picture in display order.

granules [layer 2] [audio] In the MPEG specification, the set of 3 consecutive subband samples from all 32 subbands that are considered together before quantization. They correspond to 96 PCM samples.

granules [layer 3] [audio] 576 frequency lines that carry their own side information.

group of pictures [video] A series of one or more coded pictures intended to assist random access.

GUI Graphic User Interface; for example, Microsoft Windows, Macintosh System 7, OS/2, etc.

HCR Abbreviation for Hodge Computer Research Corporation.

HDTV High Definition Television, generally compatible with the philosophies in this book.

head end Originating point of a signal in a video transmission system. In a video dial tone system, the head end is where content sent by a number of information providers is combined into a single signal and sent out via fiber optics to neighborhoods. It is the location at which television program source material is collected by direct reception, via satellite, from local sources directly or by upstream contribution links, and by various means of storage. This source material and other signals are assembled in frequency-division multiplex (FDM) in an RF band, and this signal is launched through the trunk and feeder systems "downstream" toward customers. Head ends may serve a few hundreds or thousands of subscribers in small localities, and tens or hundreds of thousands in larger localities.

header A block of data in the coded bit stream containing the coded representation of a number of data elements pertaining to the coded data that follow the header in the bit stream.

Host Digital Terminal (HDT) The HDT acts as a collection point for all the information from the network interface units and the distribution plant. The HDT electronically concentrates and manages traffic, and serves as the interconnection point into the local switching system (such as a DMS or 5ESS).

Huffman coding A specific method for entropy coding.

hybrid filter bank [audio] A serial combination of subband filter bank and MDCT.

hybrid scalability Hybrid scalability is the combination of two (or more) types of scalability.

I-field picture A field-structure I-picture.

I-frame picture A frame-structure I-picture.

I-picture An intra-coded picture; a picture coded using information only from itself.

IBM International Business Machines Corporation.

IEEE Institute of Electrical & Electronic Engineers.

IMDCT [audio] Inverse Modified Discrete Cosine Transform.

Information Superhighway Also known as the National Information Infrastructure (NII). It is intended to be the ultimate wide area network, carrying data at multiple gigabytes per second. When the CATV and TELCOs connect their multi-gigabyte ITV and advanced multimedia networks on a national basis, this will become the defacto NII. The information superhighway will probably grow to become a worldwide system.

intensity stereo [audio] A method of exploiting stereo irrelevance or redundancy in stereophonic audio programs based on retaining at high frequencies only the energy envelope of the right and left channels.

interlace The property of conventional television frames where alternating lines of the frame represent different instances in time. In an interlaced frame, one of the fields is meant to be displayed first. This field is called the first field. The first field can be the top field or the bottom field of the frame.

interactive communication The opposite of one-way or passive communication. Users of interactive services have the ability to actively respond to and control the information they receive. For example, users of an interactive news service might be able to stop a news program briefly to focus on a story that is of particular interest. They might ask to review previous stories on the same topic, view additional taped footage that didn't fit into the traditional news time slot, or see printed stories that go into more detail. Another example of interactive services might be a video game where players pit their skills against those of other players on line at the same time.

interlace [video] The property of conventional television pictures where alternating lines of the picture represent different instances in time.

intra coding Coding of a macroblock or picture that uses information only from that macroblock or picture.

ISDN (Integrated Services Digital Network) This technology maximizes the transmission capability of exiting copper wires, allowing for the simultaneous digital transmission of voice, data, and video signals over a pair of twisted wires. ISDN makes possible a wide variety of useful business and consumer applications, such as telecommuting, video teleconferencing, screen sharing, large file transfer (used in printing and publishing applications), high-speed access to remote databases, and high-speed facsimile. Greater bandwidth is needed, however, for full-motion video.

ISO/IEC 11172 (multiplexed) stream [system] A bit stream composed of one or more elementary streams combined in the manner defined in ISO/IEC 11172-1.

ITV Abbreviation for Interactive Television.

joint stereo coding [audio] Any method that exploits stereophonic irrelevance or stereophonic redundancy.

joint stereo mode [audio] A mode of the audio coding algorithm using joint stereo coding.

LAN Local Area Network, as opposed to WAN (Wide Area Network).

layer [audio] One of the levels in the coding hierarchy of the audio systems.

layer [video and systems] One of the levels in the data hierarchy of the video and system specifications defined in ISO/IEC 11172-1 and ISO/IEC 11172-2.

low-frequency enhancement channel [audio] A limited bandwidth channel for low frequency audio effects in a multichannel system.

luminance (component) [video] A matrix, block, or single pixel representing a monochrome representation of the signal and related to the primary colors in the manner defined in CCIR Rec 601. The symbol used for luminance is Y.

interoperability The ability of different networks, video servers, and set top boxes to operate compatibly together.

level A defined set of constraints on the values that may be taken by the system's parameters within a particular profile. A profile may contain one or more levels.

luminance (component) A matrix, block, or single sample representing a monochrome representation of the signal and related to the primary colors in the manner defined in the bit stream. The symbol used for luminance is Y.

macroblock The four 8×8 blocks of luminance data and the two (for 420 chrominance format), four (for 422 chrominance format) or eight (for 444 chrominance format) corresponding 8×8 blocks of chrominance data coming from a 16×16 section of the luminance component of the picture. Macroblock is sometimes used to refer to the sample data, and sometimes for the coded representation of the sample values and other data elements defined in the macroblock header of the syntax.

Mann window [audio] A time function applied sample-by-sample to a block of audio samples before Fourier transformation.

mapping [audio] Conversion of an audio signal from time to frequency domain by subband filtering and/or by MDCT.

masking [audio] A property of the human auditory system by which an audio signal cannot be perceived in the presence of another audio signal.

masking threshold [audio] A function in frequency and time below which an audio signal cannot be perceived by the human auditory system.

MDCT [audio] Modified Discrete Cosine Transform, which corresponds to the Time Domain Aliasing Cancellation Filter Bank.

measured service A service for which charges are levied based on actual usage (rather than availability).

motion compensation [video] The use of motion vectors to improve the efficiency of the prediction of pixel values. The prediction uses motion vectors to provide offsets into the past and/or future reference pictures containing previously decoded pixel values that are used to form the prediction error signal.

motion estimation [video] The process of estimating motion vectors during the encoding process.

motion vector A two-dimensional vector used for motion compensation that provides an offset from the coordinate position in the current picture or field to the coordinates in a reference frame or reference field.

movies on demand A service that allows customers to (a) select the movie they'd like to see from an extensive library of past and present movies, and (b) select the time that movie is played on their home screen (see near movies on demand).

MS stereo [audio] A method of exploiting stereo irrelevance or redundancy in stereophonic audio programs based on coding the sum and difference signal instead of the left and right channels.

MSO Multiple Service Operator.

multichannel [audio] A combination of audio channels used to create a spatial sound field.

multilingual [audio] A presentation of dialogue in more than one language.

multimedia The use of multiple forms of media to communicate; i.e., audio, video, text, graphics, etc. The combination of these media allows for more powerful and effective communication.

multimode Multiple optical transmission modes in a fiber.

multiplexing Process combining channels to aggregates or groups.

multipoint Having more than two points of transmission and/or reception.

multipoint distribution service A one-way domestic public radio service on microwave frequencies, from a fixed station transmitting (usually in an omnidirectional pattern) to multiple receiving facilities located at fixed points determined by subscribers.

narrowband network See broadband network.

National Information Infrastructure See Information Superhighway.

NCTA Abbreviation for National Cable Television Association.

near video on demand Service where customers are given a limited choice of videos to view with published viewing times. (For example, the 20 most popular new releases at any viewing time might be made available, and a particular selection might be aired every 5 minutes.)

neighborhood node The location in a neighborhood where a video signal (sent from the head end) is converted from an optical signal to an electronic signal. In other words, the node replaces the information it receives over the fiber-optic cable, and resends it in a new format over coaxial cable to the home.

network assurance The network should be built with intelligent elements capable of detecting and isolating faults at any point in the delivery chain. Continuous service monitoring should be performed from the head end to the side of the customer's home. Today, no other telephony network or cable TV network has such a complete service assurance capability.

network interface unit (NIU) A device attached to the side of the user's house that acts as a network interface for telephone service, ISDN, and broadband consumer services. The NIU manages these services and continuously monitors the network to ensure information is properly relayed to the customer's residence. It converts analog signals on the phone into digital signals and sends them into the network for processing.

network provisioning Services in the network should be software-controlled. The design will permit the head end to change customer services by changing software rather than rearranging physical circuits. This reduces installation times and operating expenses. And because the network elements are intelligent, it is possible to reduce both database and data management activities. The network should be capable of self-inventory and self-reporting as well.

nonintra-coding Coding of a macroblock or picture that uses information both from itself and from macroblocks and pictures occurring at other times.

nontonal component [audio] A noiselike component of an audio signal.

NUI Network Unit Interface.

Nyquist sampling Sampling at or above twice the maximum bandwidth of a signal.

OCR Optical character recognition, the ability of a computer system to automatically recognize letters in printed text and convert them into word processor-compatible data.

OS Abbreviation for operating system.

P-field picture A field structure P-picture.

P-frame picture A frame structure P-picture.

P-picture; predictive-coded picture A picture that is coded using motion compensated prediction from past reference fields or frame.

packet data [system] Contiguous bytes of data from an elementary stream present in a packet.

packet header [system] The data structure used to convey information about the elementary stream data contained in the packet data.

packet [system] A packet consists of a header followed by a number of contiguous bytes from an elementary data stream.

padding [audio] A method to adjust the average length in time of an audio frame to the duration of the corresponding PCM samples, by conditionally adding a slot to the audio frame.

PANS Pretty Amazing New Services.

PPV Pay-per-view programming where the subscriber pays only for the programs he watches.

parameter A variable within the syntax of a specification that may take one of a large range of values. A variable that can take one of only two values is a flag and not a parameter.

past reference frame (field) A past reference frame (field) is a reference frame (field) that occurs at an earlier time than the current picture in display order.

past reference picture [video] The past reference picture is the reference picture that occurs at an earlier time than the current picture in display order.

pay-per-view Programs available to a subscriber on a pay-as-you-watch basis.

personal communication services (PCS) A broad range of individualized wireless telecommunications services that let people or devices communicate irrespective of their physical location.

picture Source, coded, or reconstructed image data. For example, an MPEG source or reconstructed picture consists of three rectangular matrices of 8-bit numbers representing the luminance and two chrominance signals. For progressive video, a picture is identical to a frame, while for interlaced video, a picture can refer to a frame, the top field of the frame, or the bottom field of the frame depending on the context.

picture period [video] The reciprocal of the picture rate.

picture rate [video] The nominal rate at which pictures are output from the decoding process.

pixel aspect ratio [video] The ratio of the nominal vertical height of pixel on the display to its nominal horizontal width.

pixel [video] Picture element.

polyphase filter bank [audio] A set of equal bandwidth filters with special phase interrelationships, allowing for an efficient implementation of the filter bank.

POTS Plain Old Telephone Service.

power It is desirable to continue to provide backup network power in the event of commercial power failure. Using a single network, backup power can also support broadband network services.

prediction The use of a predictor to provide an estimate of the sample value or data element currently being decoded.

prediction [audio] The use of a predictor to provide an estimate of the subband sample in one channel from the subband samples in other channels.

prediction error The difference between the actual value of a sample or data element and its predictor.

predictor A linear combination of previously decoded sample values or data elements.

presentation channel [audio] Audio channels at the output of the decoder corresponding to the loudspeaker positions left, center, right, left surround, and right surround.

presentation time stamp (PTS) [system] A field that may be present in a packet header that indicates the time at which a presentation unit is presented in the system target decoder.

progressive The property of film frames where all the samples of the frame represent the same instances in time.

PROM Programmable Read-Only Memory.

quality of service By extending digital technology from the head end to the side of the customer's home, significant improvements are made in the quality of the information services received by the customer. Voice services will no longer be affected by noise and loss encountered in today's outside plant. Data services can be delivered at Ethernet rates to the side of the home using digital encoding technology. Broadband signal quality can be kept well above customer perception levels using fiber-optic technology and short coaxial distribution runs.

quantization matrix In MPEG, a set of 64 8-bit values used by the dequantizer.

quantized DCT coefficients DCT coefficients before dequantization. A variable-length coded representation of quantized DCT coefficients is transmitted as part of the compressed-video bit stream.

quantizer scale A scale factor coded in the bit stream and used by the decoding process to scale the dequantization.

RAID Redundant Array of Independent Disks. A video RAID is a unit specially designed to store multiple video streams and programs.

RAM Random Access Memory.

random access The process of beginning to read and decode the coded bit stream at an arbitrary point.

reconstructed frame A reconstructed frame consists of three rectangular matrices of 8-bit numbers representing the luminance and two chrominance signals. A reconstructed frame is obtained by decoding a coded frame.

reconstructed picture A reconstructed picture is obtained by decoding a coded picture. A reconstructed picture is either a reconstructed frame (when decoding a frame picture), or one field of a reconstructed frame (when decoding a field picture). If the coded picture is a field picture, then the reconstructed picture is either the top field or the bottom field of the reconstructed frame.

reference field A reference field is one field of a reconstructed frame. Reference fields are used for forward and backward prediction when P-pictures and B-pictures are decoded. Note that when field P-pictures are decoded, prediction of the second field P-picture of a coded frame uses the first reconstructed field of the same coded frame as a reference field.

reference frame A reference frame is a reconstructed frame that was coded in the form of a coded I-frame or a coded P-frame. Reference frames are used for forward and backward prediction when P-pictures and B-pictures are decoded.

robbing Occasional stealing of information bits for signaling.

ROM Read Only Memory.

SAES Small Aperture Earth Stations (small-antenna satellite ground stations).

sample aspect ratio (SAR) This specifies the distance between samples. It is defined (for the purposes of this specification) as the vertical displacement of the lines of luminance samples in a frame divided by the horizontal displacement of the luminance samples. Thus its units are (meters per line) (meters per sample).

sampling Observation and recording (often with the intent of quantization).

scalability The ability of a decoder to decode an ordered set of bit streams to produce a reconstructed sequence. Moreover, useful video is output when subsets are decoded. The minimum subset that can thus be decoded is the first bit stream in the set which is called the base layer. Each of the other bit streams in the set is called an enhancement layer. When addressing a specific enhancement layer, "lower layer" refer to the bit stream that precedes the enhancement layer.

SCSI Small Computer System Interface, a popular disk drive interface used by many other peripheral devices.

side information Information in the bit stream necessary for controlling the decoder.

skipped macroblock A macroblock for which no data is encoded.

slice A series of macroblocks.

slot [audio] A slot is an elementary part in the bit stream. In MPEG Layer I a slot equals four bytes, in Layers II and III one byte.

SMPTE Society of Motion Picture & Television Engineers.

SNR scalability A type of scalability where the enhancement layer(s) contain only coded refinement data for the DCT coefficients of the lower layer.

SONET Synchronous Optical Data Network.

source stream A single nonmultiplexed stream of samples before compression coding.

space division Signal separation by physical distance, as with open wire lines on a pole or wires in a cable.

spatial scalability A type of scalability where an enhancement layer also uses predictions from sample data derived from a lower layer without using motion vectors. The layers can have different frame sizes, frame rates, or chrominance formats.

spreading function [audio] A function that describes the frequency spread of masking effects.

SQL System Query Language, a popularized outgrowth of IBM's DB2 query language.

start codes [system and video] 32-bit codes embedded in a coded bit stream that are unique. They are used for several purposes, including identifying some of the structures in the coding syntax.

stereo-irrelevant [audio] A portion of a stereophonic audio signal that does not contribute to spatial perception.

stuffing (bits); stuffing (bytes) Code words that may be inserted into the coded bit stream that are discarded in the decoding process. Their purpose is to increase the bit rate of the stream.

subband [audio] Subdivision of the audio frequency band.

subband filter bank [audio] A set of band filters covering the entire audio frequency range.

subband samples [audio] The subband filter bank within the audio encoder creates a filtered and subsampled representation of the input audio stream. The filtered samples are called subband samples. From 384 time-consecutive input audio samples, 12 time-consecutive subband samples are generated within each of the 32 subbands.

surround channel [audio] An audio presentation channel added to the front channels (L and R or L, R, and C) to enhance the spatial perception.

synchronous A mode of transmission in which symbols are exchanged with fixed, rather than random, intersymbol time separation.

sync word [audio] A 12-bit MPEG code embedded in the audio bit stream that identifies the start of a frame.

synthesis filter bank [audio] Filter bank in the decoder that reconstructs a PCM audio signal from subband samples.

system header [system] The system header is a data structure defined in ISO/IEC 11172-1 that carries information summarizing the system characteristics of the ISO/IEC 11172 multiplexed stream.

system target decoder; STD [system] A hypothetical reference model of a decoding process used to describe the semantics of an ISO/IEC 11172 multiplexed bit stream.

TELCO Telephone Company.

telecommunications Point-to-point electrical or electronic transmission of voice, data, and video images.

telecommunications infrastructure The underlying structure or framework of the telecommunications system; the cable, conduit, switching machines, amplifiers, and support systems that allow for the transmission of voice, video, and data.

temporal scalability A type of scalability where an enhancement layer also uses predictions from sample data derived from a lower layer using motion vectors. The layers have identical frame size and chrominance formats, but can have different frame rates.

thread a single program stored on a magnetic disk can frequently be read out many times in real time, due to its bandwidth (which is often multiples of the required video bandwidth). Each readout can represent a different time slot of the program and is designated a thread.

time division Separation on a chronological basis.

time stamp [system] A data segment attached to a packet or file that indicates the time of an event.

triplet [audio] A set of 3 consecutive subband samples from one subband. A triplet from each of the 32 subbands forms a granule.

tonal component [audio] A component of an audio signal, usually sinusoidal.

top field One of two fields that comprise a frame. Each line of a top field is spatially located immediately above the corresponding line of the bottom field.

transponder A frequency translating amplifier in a communications satellite.

trunk cable Coaxial cables that carry the RF signal from the head end to groups of subscribers. Amplitude-modulated wideband microwave systems are sometimes used to bridge obstacles or large distances. Losses in the trunk coaxial cable are made up by trunk amplifiers, which are designed to balance degradations due to noise and distortion. Fiber-to-the-serving-area (FSA) designs now limit the number of active amplifiers after a fiber node to a range of four to ten.

TV Television.

TVOD True Video On Demand. It provides the subscriber with a movie instantaneously, as compared to NVOD, which, although it provides an instantaneous response, may take seconds or minutes before the requested video begins.

underflow Overdepletion of bits in a buffer.

uplink Signal path from the earth to satellites.

upstream Towards the head end, a reverse signaling path.

variable bit rate Operation where the bit rate varies with time during the decoding of a coded bit stream.

variable length coding (VLC) A reversible procedure for coding that assigns shorter code-words to frequent events and longer code-words to less frequent events.

video buffering verifier (VBV) A hypothetical decoder that is conceptually connected to the output of the encoder. Its purpose is to provide a constraint on the variability of the data rate that an encoder or editing process may produce.

video dial tone An FCC-authorized service in which participating local telephone companies are required to carry, without discrimination, the video programming of all requesting parties.

video dial tone gateway The video dial tone gateway will act as a switching point in the central office for the delivery of video and multimedia services to the home.

video file server A device capable of storing video information in a compressed digital format; multiple users can simultaneously access that video information on a random basis, creating a VCR-like capability that can be shared over a wide area.

video on demand VOD. Video available to a cable subscriber when he or she desires it, approximately when he or she wants it.

video sequence The highest syntactic structure of coded video bit streams. It contains a series of one or more coded frames. It is one of the layers of the coding syntax defined in ISO/IEC 11172-2.

virtual reality The stimulation of most of the human senses via advanced multimedia technology, so as to produce the illusion of an alternative reality.

voice telephone service The ability to place conventional telephone service digitally on the coaxial network concurrent with ITV, advanced multimedia, and other data services.

VSAT Very Small Aperture Terminal (a small satellite earth station).

WAN Wide Area Network as opposed to LAN (Local Area Network). The NII is a WAN.

wideband Bandwidth equal to many individual channels.

wideband network See broadband.

zig-zag scanning order A specific sequential ordering of the DCT coefficients from (approximately) the lowest spatial frequency to the highest.

Bibliography

Anderson, Gary H. *Video Editing & Post Production.* Knowledge Industry Publications, Inc. New York, 1988.

Andrews, Harry C. *Digital Image Processing.* IEEE Computer Society. New York, 1978.

Avedon and Levey. *Electronic Imaging Systems.* McGraw-Hill, Inc. New York, 1991.

Ball, Larry L. *Network Management with Smart Systems.* McGraw-Hill, Inc. New York, 1994.

Benson and Fink. *HDTV: Advanced Television for the 1990s.* McGraw-Hill, Inc. New York, 1991.

Benson and Whitaker. *Television Engineering Handbook.* McGraw-Hill, Inc. New York, 1992.

Black, Uyless. *TCP/IP and Related Protocols.* McGraw-Hill, Inc. New York, 1992.

Bogner and Constantinides. "Introduction to Digital Filtering." John Wiley & Sons. Chichester, 1975.

Chiddix, James A. "Fiber Backbone—Multi-Channel AM Video Trunking." *Proceedings of the Montreaux International Television Symposium.* June 1989.

Chiddix, James A. "Fiber Backbone for Cable TV Using Multi-Channel AM Video Trunking." *International Journal of Digital and Analog Cabled Systems.* 1989.

Chiddix, James A. "Application of Optical Fiber Transmission Technology to Existing CATV Networks." *Proceedings of 12th Annual International Fiber Optic Communication and Local Area Network Conference and Exposition.* September 1988.

Chiddix, James A. "Fiber Optic Supertrunking." *Communications Engineering and Design.* September 1985.

Chiddix, James A. "Fiber Optic Technology for CATV Supertrunk Application." *NCTA Technical Papers.* 1985.

Chiddix, James A. "Introduction of Optical Fiber Transmission Technology into Existing Cable Television Networks and Its Impact on the Consumer Electronics interface." *IEEE Transactions on Consumer Electronics,* Vol. 96. No. 2. May 1989.

Chiddix, James A. "Optical Fiber Supertrunking. The Time Has Come; A Performance Report On A Real-World System." *IEEE Journal on Selected Areas In Communications.* Vol. SAC-4. August 1986.

Chiddix and Pangrac. "Fiber Backbone—Multi-Channel AM Video Trunking." *NCTA Technical Papers.* 1989.

Chiddix and Pangrac. "Fiber Backbone: A Proposal for An Evolutionary CATV Architecture." *NCTA Technical Papers.* October 1988.

Chiddix and Pangrac. "Off Premises Broadband Addressability: A CATV Industry Challenge." *NCTA Technical Papers.* 1989.

Dudgeon and Mersereau. *Multidimensional Digital Signal Processing.* Prentice Hall. New Jersey, 1984.

Fischler and Firschein. *Computer Vision.* Morgan Kaufmann Publishers, Inc, California.

Gopal, Inder. *Planet Orbit Projects.* IBM. New York, 1994.

Grant and Leavenworth. *Statistical Quality Control.* McGraw-Hill, Inc. New York, 1988.

Higgins, Richard J. *Digital Signal Processing.* Prentice Hall. New Jersey, 1990.

Hodge, Block, and Harvey. "A Film Quality Digital Archiving & Editing System." *SMPTE: Advanced Television and Electronic Imaging for Film & Video,* 57-73; New York, 1993.

Hodge, Mabon, and Powers. "Video On Demand: Architecture. Systems & Applications." *SMPTE Journal.* Volume 102. Number 9. New York. September 1993.

Hodge and Milligan. "TVOD vs. NVOD: Statistical Modeling. Cost & Performance Trade-Offs." *National Cable Television Association—Technical Papers.* 157-172. New Orleans, 1994.

Inglis, Andrew F. *Video Engineering.* McGraw-Hill, Inc. New York, 1993.

Intercor Publications. "Inter Active Television Reports." Intercor Publications. September 1992 through August 1994, inclusive.

International Business Machines Corporation. *IBM MPEG-2 Decoder Chip User's Guide* (DVP4SCUMU-02). IBM. New York, 1994.

Jayant and Noll. *Digital Coding of Waveforms.* Prentice-Hall. New Jersey, 1984.

Jones, Edwin. *Digital Transmission.* McGraw-Hill, Inc. England, 1993.

Korth and Silberschatz. *Database System Concepts.* McGraw-Hill, Inc. New York, 1991.

Larijani, L. Casey. *The Virtual Reality Primer.* McGraw-Hill, Inc. New York, 1994.

Lim, Jae S. *Two Dimensional Signal & Image Processing.* Prentice Hall. New Jersey, 1990.

Luther, Arch C. *Digital Video in the PC Environment.* McGraw-Hill, Inc. New York, 1991.

MPEG 2 Committee. "Generic Coding of Moving Pictures and Associated Audio Recommendation H.222.0." International Organization for Standardization (ISO/IEC 13818). Washington D.C., 24 June 1994.

Nemzow, Martin A. W. *FDDI Networking.* McGraw-Hill, Inc. New York, 1993.

Pearson, Don. *Image Processing.* McGraw-Hill, Inc. England, 1991.

Pelton, Gordon E. *Voice Processing.* McGraw-Hill, Inc. New York, 1993.

Powers and Stair. *Megabit Data Communications.* Prentice Hall. New Jersey, 1990.

de Prycker, Martin. *Asynchronous Transfer Mode.* Ellis Harwood, New York, 1993.

Rishe, Naphtali. *Database Design.* McGraw-Hill, Inc. New York, 1992.

Russ, John C. *The Image Processing Handbook.* CRC Press. Boca Raton, 1992.

Spohn, Darren L. *Data Network Design.* McGraw-Hill, Inc. New York, 1993.

Stafford, R. H. *Digital Television.* John Wiley & Sons. New York, 1980.

Texas Instruments. "ATM and SONET—Broadband Solutions for LAN and WAN Applications." Texas Instruments, USA, 1994

Winch, Robert G. *Telecommunications Transmission Systems.* McGraw-Hill, Inc. New York, 1993.

Index

T

U

V

ABOUT THE AUTHOR

Win Hodge, founder and chief executive of HCR and founder of IPC, is a versatile executive who has been orchestrating the deployment of advanced-technology computer architectures, video servers and imaging algorithms for image enhancement, compression, identification and tracking, this for a period exceeding 30 years. He has numerous high-technology publications, patents, inventions, and mass-produced products. His products have produced over six billion dollars for his clients.

In the mid 1960s he headed up the Earth Resource Technology Satellite Program for Rockwell International, which was a project almost entirely consisting of imaging, signal processing, compression, computer imaging architecture, simulation, image databasing, etc. In the mid 1970s he was responsible for the Hotel Pay Per View Video On Demand system manufactured by Paramount-Summit (Spectradyne). Subsequently, he was involved with the Select TV over-the-air broadcast Pay Per View system. In the 1980s he applied advanced multimedia and image database technology from projects ranging from TV to newspaper publishing systems to aircraft simulators. In the 1990s he has been focusing on video servers, ATM switches, and the set-top boxes for interactive TV applications.

He has been involved with imaging, computer architecture, and artificial intelligence since 1962 (32 years of directly related experience). He is a member of the Society of Motion Picture and Television Engineers (SMPTE) and the IEEE and its IEEE computer, imaging, communications, and artificial intelligence societies.

Education

Mr. Hodge graduated from Chapman University in 1962, where he received his bachelor's degree majoring in physics and mathematics. His graduate work at UCLA & CSUF was in statistical communications theory.

Clients

Win Hodge's clients have included Eastman Kodak, Digital Equipment Corporation, The Silicon Valley Group, Intel, McDonnell Douglas, Memorex, National Semiconductor, Reliance, Rockwell International, Storage Technology Corporation, Sunbeam, TRW, Telefunken, Tektronix, Western Digital, and the United States Government.